Design of Survivable Networks with Bounded Rings

Network Theory and Applications

Volume 2

Design of Survivable Networks with Bounded Rings

by

Bernard Fortz

Université Libre de Bruxelles

SPRINGER-SCIENCE+BUSINESS MEDIA, B.V.

A C.I.P. Catalogue record for this book is available from the Library of Congress.

ISBN 978-1-4613-7113-7 ISBN 978-1-4615-4669-6 (eBook)
DOI 10.1007/978-1-4615-4669-6

Printed on acid-free paper

To Sophie and Benoît

Contents

List of Figures xi
List of Tables xiii
List of Algorithms xvii
Acknowledgments xix

1. INTRODUCTION 1

2. SURVIVABLE NETWORK DESIGN: A SURVEY 5
   - 2.1 Notation and definitions 6
   - 2.2 Low-connectivity constrained network design problems 8
   - 2.3 Structural properties and particular cases 12
   - 2.4 Heuristics 17
   - 2.5 Polyhedral studies and exact algorithms 21

3. TWO-CONNECTED NETWORKS WITH BOUNDED RINGS: THE MODEL 25
   - 3.1 Motivation 26
   - 3.2 Mathematical formulations 27

4. POLYHEDRAL STUDY 35
   - 4.1 Associated polytopes and trivial inequalities 36
   - 4.2 Cut constraints 40
   - 4.3 Node-cut and subset inequalities 46
   - 4.4 Ring-cut inequalities 51
   - 4.5 Ring-cover inequalities 58
   - 4.6 Metric inequalities 59

4.7 Node-partition inequalities 63
4.8 Weighted partition inequalities 65

5. THE SPECIAL CASE OF RINGS WITH BOUNDED CARDINALITY 69
5.1 A lower bound on the number of edges 70
5.2 Complexity 77
5.3 Cyclomatic inequalities 81
5.4 Triangular rings 88

6. A BRANCH-AND-CUT ALGORITHM 91
6.1 Checking feasibility 92
6.2 Separation of cut constraints 95
6.3 Separation of metric inequalities 95
6.4 Separation of subset inequalities 100
6.5 Separation of ring-cut inequalities 102
6.6 Separation of node-partition inequalities 106
6.7 Separation of weighted partition inequalities 106
6.8 Separation of cyclomatic inequalities 107
6.9 Implementation of the Branch-and-Cut algorithm 107
6.10 Summary of polyhedral results 113

7. HEURISTICS 115
7.1 Ear-inserting method 116
7.2 Cutting cycles in two equal parts 116
7.3 Path-following method 117
7.4 Stingy method 119
7.5 Tabu Search 119

8. COMPUTATIONAL RESULTS 125
8.1 Branch-and-Cut for Euclidean edge lengths 127
8.2 Branch-and-Cut for unit edge lengths 133
8.3 Heuristics for Euclidean edge lengths 138
8.4 Heuristics for unit edge lengths 143

9. CONCLUSION 147

Appendices 149
A– Detailed Computational Results 149
A.1 Description of the test problems 150

A.2 Branch-and-Cut for Euclidean edge lengths 158
A.3 Branch-and-Cut for unit edge lengths 165
A.4 Heuristics for Euclidean edge lengths 172
A.5 Heuristics for unit edge lengths 185

References 193

Index 201

List of Figures

3.1	Minimum cost two-connected network	28
4.1	Example of graph belonging to $\mathcal{D}om\mathcal{P}_{G,K}$ but not to $\mathcal{P}_{G,K}$	37
4.2	An example for which cut constraints are not facet-defining ($K = 4$)	41
4.3	Ring-cut graph for $W = \{1,2,3,4,5,6,7\}$ ($K = 4$)	43
4.4	$G - (\delta(W)\backslash\delta(\{z\}))$	47
4.5	A graph such that $G - S$ is not two-connected and $S \neq \delta(W)\backslash\delta(\{z\})$	51
4.6	Graph for which some ring-cut inequalities are valid	52
4.7	The ring-cut graph for $W = \{1,2,3,4,5,6\}$	53
5.1	An example of graph and its fundamental cycle matrix	71
5.2	An illustration of Theorem 5.1	73
5.3	Combinations of a cycle σ and an outside path P to generate cycles σ_1 and σ_2	76
5.4	Transformation of Hamiltonian cycle into 2CNBR	80
5.5	A solution of R2CNBR and the corresponding Hamiltonian cycle	81
5.6	A graph G, the contracted graph $\hat{G}$ and the two steps of $\overline{G}$ construction, with a bound $K = 4$.	83
5.7	F_1 and F_2	85
5.8	F_3 and F_4	87
6.1	Values of $\alpha_k = \pi_k$	97
6.2	Transformation of independent subset into RING-CUT-SEP	103
6.3	Approximation of $M(p, 4)$ for $n = 50$	108

6.4	Approximation of $M(p, 7)$ for $n = 50$	109
8.1	Optimal solution, 30 nodes real instance, $K = 200$	128
8.2	Optimal solution, 30 nodes real instance, $K = 350$	128
8.3	Optimal solution, 30 nodes real instance, $K = 6$	134
8.4	Optimal solution, 30 nodes real instance, $K = 8$	134

List of Tables

6.1	A summary of valid inequalities	114
8.1	List of abbreviations	129
8.2	Branch-and-Cut results, real applications, Euclidean edge lengths	130
8.3	Inequalities generated for the 52 nodes real instance, Euclidean edge lengths	131
8.4	Branch-and-Cut results, random networks, Euclidean edge lengths	132
8.5	Branch-and-Cut results, real applications, unit edge lengths	135
8.6	Inequalities generated for the 52 nodes real instance, unit edge lengths	136
8.7	Branch-and-Cut results, random networks, unit edge lengths	137
8.8	Constructive heuristics, real applications, Euclidean edge lengths	139
8.9	Constructive heuristics, random networks, Euclidean edge lengths	140
8.10	Tabu Search, real applications, Euclidean edge lengths	141
8.11	Tabu Search, random networks, Euclidean edge lengths	142
8.12	Tabu Search, real applications, unit edge lengths	144
8.13	Tabu Search, random networks, unit edge lengths	145
A.1	Real applications, Euclidean edge lengths	151
A.2	Random networks (10 nodes), Euclidean edge lengths	152
A.3	Random networks (20 nodes), Euclidean edge lengths	153
A.4	Random networks (30 nodes), Euclidean edge lengths	154
A.5	Random networks (40 nodes), Euclidean edge lengths	155
A.6	Random networks (50 nodes), Euclidean edge lengths	156
A.7	Instances with unit edge lengths	157

A.8 Branch-and-Cut results, real applications, Euclidean edge lengths 159

A.9 Branch-and-Cut results, random networks (10 nodes), Euclidean edge lengths 160

A.10 Branch-and-Cut results, random networks (20 nodes), Euclidean edge lengths 161

A.11 Branch-and-Cut results, random networks (30 nodes), Euclidean edge lengths 162

A.12 Branch-and-Cut results, random networks (40 nodes), Euclidean edge lengths 163

A.13 Branch-and-Cut results, random networks (50 nodes), Euclidean edge lengths 164

A.14 Branch-and-Cut results, real applications, unit edge lengths 166

A.15 Branch-and-Cut results, random networks (10 nodes), unit edge lengths 167

A.16 Branch-and-Cut results, random networks (20 nodes), unit edge lengths 168

A.17 Branch-and-Cut results, random networks (30 nodes), unit edge lengths 169

A.18 Branch-and-Cut results, random networks (40 nodes), unit edge lengths 170

A.19 Branch-and-Cut results, random networks (50 nodes), unit edge lengths 171

A.20 Constructive heuristics, real applications, Euclidean edge lengths 173

A.21 Constructive heuristics, random networks (10 nodes), Euclidean edge lengths 174

A.22 Constructive heuristics, random networks (20 nodes), Euclidean edge lengths 175

A.23 Constructive heuristics, random networks (30 nodes), Euclidean edge lengths 176

A.24 Constructive heuristics, random networks (40 nodes), Euclidean edge lengths 177

A.25 Constructive heuristics, random networks (50 nodes), Euclidean edge lengths 178

A.26 Tabu Search, real applications, Euclidean edge lengths 179

A.27 Tabu Search, random networks (10 nodes), Euclidean edge lengths 180

A.28 Tabu Search, random networks (20 nodes), Euclidean edge lengths 181

A.29 Tabu Search, random networks (30 nodes), Euclidean edge lengths 182
A.30 Tabu Search, random networks (40 nodes), Euclidean edge lengths 183
A.31 Tabu Search, random networks (50 nodes), Euclidean edge lengths 184
A.32 Tabu Search, real applications, unit edge lengths 186
A.33 Tabu Search, random networks (10 nodes), unit edge lengths 187
A.34 Tabu Search, random networks (20 nodes), unit edge lengths 188
A.35 Tabu Search, random networks (30 nodes), unit edge lengths 189
A.36 Tabu Search, random networks (40 nodes), unit edge lengths 190
A.37 Tabu Search, random networks (50 nodes), unit edge lengths 191

List of Algorithms

6.1 Generic depth-first search of an undirected graph $G = (V, E)$ 93
6.2 The function $check\_cycle(G, e, ok, inf, num)$ 95
6.3 Checking two-connectivity of G_K 96
6.4 Separation of metric inequalities 99
6.5 Separation of subset inequalities 101
6.6 Separation of ring-cut inequalities 105
6.7 Branch-and-Cut 110
7.1 Ear-inserting heuristic 117
7.2 Cut-in-two heuristic 118
7.3 Stingy heuristic 119
7.4 Descent method 120
7.5 Tabu Search heuristic 123

Acknowledgments

This book is the result of about six years of work, and it would not have been possible without the constant help and advice of my thesis supervisor Martine Labbé.

During these years, I spent a few months in Milan, and I would like to thank Francesco Maffioli for this collaboration and Alessandro Segalini for the help provided in implementing heuristics.

Many people have contributed in improving this text with their useful suggestions and corrections: thanks to Auguste Caverenne, Jan Karel Lenstra, Georges Lognay, Paolo Toth and Laurence Wolsey for the precious time they spent reading this work.

Finally, my wife Florence constantly supported me and I want to thank her for her patience and encouragements.

Chapter 1

INTRODUCTION

These days, the nature of services and the volume of demand in the telecommunication industry is changing radically, with the replacement of analog transmission and traditional copper cables by digital technology and fiber optic transmission equipment. Moreover, we see an increasing competition among providers of telecommunication services, and the development of a broad range of new services for users, combining voice, data, graphics and video. Telecommunication network planning has thus become an important problem area for developing and applying optimization models.

Telephone companies have initiated extensive modeling and planning efforts to expand and upgrade their transmission facilities, which are, for most national telecommunication networks, divided in three main levels (see Balakrishnan et al. [5]), namely,

1. the *long-distance* or *backbone* network that typically connects city pairs through *gateway nodes*;

2. the *inter-office* or *switching center* network within each city, that interconnects *switching centers* in different subdivisions (clusters of customers) and provides access to the gateway(s) node(s);

3. the *local access* network that connects individual subscribers belonging to a cluster to the corresponding switching center.

These three levels differ in several ways including their design criteria. Ideally, the design of a telecommunication network should simultaneously account for these three levels. However, to simplify the planning task, the overall planning problem is decomposed by considering each level separately.

This work is the result of a project funded by Belgacom, the Belgian phone operator, dealing with the development of new models and techniques for *long-term planning* in the first level of the hierarchy, i.e. the *backbone network*.

Planning in the backbone network is divided in two different stages : mid-term and long-term planning. Mid-term planning, covering a period of about five years, consists in dimensioning the network. More precisely, given a forecast of the demand matrix for this period and the current topology of the network, we have to compute how the expected demands will be routed as well as the necessary capacities of the cables. In some models, the addition of new edges is allowed. These problems involve, at the same time, survivable design criteria and routing constraints. A recent survey on these models can be found in De Jongh [27].

Long-term planning usually involves a ten-year timescale. Demand data for such a long period are not reliable enough, and we only deal with topological aspects. The goal is then to determine a set of cables connecting all nodes under some survivability criteria. In this context, the telephone network is seen as a given set of nodes and a set of possible fiber links that have to be placed between these nodes to achieve connectivity and survivability at minimum cost.

In traditional backbone networks, the limited capacity of copper cables resulted in highly diverse routing between offices. The developments in fiber-optic technology have led to components that are cheap and reliable, having an almost unlimited capacity. The introduction of such a technology has made hierarchical routing and bundling of traffic very attractive. This approach has resulted in sparse, even treelike network topologies with larger amounts of traffic carried by each link.

Two main issues appear in the planning process of fiber-optic networks : economy and survivability. Economy refers to the construction cost, which is expressed as the sum of the edge costs, while survivability refers to the restoration of services in the event of node or link failure. Trees satisfy the primary goal of minimizing the total cost while connecting all nodes. However, only one node or edge breakdown causes a tree network to fail in its main objective of enabling communication between all pairs of nodes.

This means that some survivability constraints have to be considered while building the network. Losing end-to-end customer service could lead to dramatic loss of revenue for commercial providers of telecommunication services. Constructing network topologies that provide protection against cable or office failures has become one of the most important problems in the field of telecommunications network design.

The most studied models deal with k-connectivity requirements, i.e. the ability to restore network service in the event of a failure of at most $k - 1$ components of the network. Among them, the minimum-cost two-connected spanning network problem consists in finding a network with minimal total cost for which two node-disjoint paths are available between every pair of nodes. This means that two-connected networks are able to deal with a single link or node failure. Two-connected networks have been found to provide a sufficient level of survivability in most cases, and a considerable amount of research has focused on so-called *low-connectivity constrained* network design problems, i.e. problems for which each node j is characterized by a requirement $r_j \in \{0, 1, 2\}$ and $\min\{r_i, r_j\}$ node-disjoint paths between every pair of nodes i, j are required. The next chapter contains a survey of the literature on these models.

When we started this project with Belgacom, we first looked at existing models, and it seemed that two-connectivity was a sufficient level of survivability for the Belgian backbone network, since the probability of dealing with two simultaneous failures is very low. We implemented a Branch-and-Cut algorithm, based on the work of Grötschel et al. [48], for finding a minimum cost two-connected network. It turned out that the optimal solution of this problem on the set of 52 zonal centers of the Belgian backbone network is a Hamiltonian cycle. Hence, any edge failure implies that the flow that passed through that edge must be rerouted, using all the edges of the network, an obviously undesirable feature. This led us to examine a new model for limiting the region of influence of the traffic which it is necessary to reroute : the *Two-Connected Network with Bounded Rings problem*

(2CNBR) is described in Chapter 2, together with some of its possible mathematical formulations. In addition to the classical two-connectivity constraints, we require in this model that each edge belongs to at least one cycle (or *ring*) whose length is bounded by a given constant. To our knowledge, this extension has not been considered before. It also finds its motivation in the emerging technology of *self-healing rings.* These are cycles in the network equipped in such a way that any link failure in the ring is automatically detected by the link end nodes and the traffic rerouted along the alternative path in the cycle. When such a strategy is chosen, rings must cover the network and their size must be limited. These two requirements are provided by our model.

From the work of Grötschel et al. [48] and Stoer [75], it turns out that the most effective formulations of the two-connected network problem as an integer linear program use an exponential number of constraints. Hence, the use of a cutting plane method is a natural approach for solving it, and has proved efficient. We adapted this method to our problem by deriving new classes of valid inequalities. These polyhedral investigations are contained in Chapter 3, leading to a Branch-and-Cut algorithm for solving the 2CNBR problem. We investigate in this chapter inequalities coming from the standard two-connected network problem as well as inequalities specific to our model. Some of these are shown to be facet-inducing.

In the case where edge lengths are equal to one, i.e. the cardinality of the rings is bounded, more structural properties and polyhedral results were found. In particular, we were able to derive a lower bound on the number of edges in any feasible solution. This result was extensively used to derive new complexity and facet results as well as new classes of inequalities that provide dramatic improvements of the lower bounds obtained by the Branch-and-Cut algorithm. Chapter 4 is devoted to this particular case.

Chapter 5 describes in detail the implementation of the Branch-and-Cut algorithm, and in particular how the separation problems are solved.

Since the problem is NP-hard, the size of the instances that can be solved exactly is limited. We thus developed heuristics that are described in Chapter 6. For some of them, we provide a worst-case performance guarantee. Computational results appear in Chapter 7.

Chapter 2

SURVIVABLE NETWORK DESIGN: A SURVEY

In this chapter, we present the most studied survivability model used for the long-term design of a backbone network, namely the low-connectivity constrained network design problem. This model is presented in Section 2.2 using graph-theoretic concepts. An integer linear programming formulation is also given. In Section 2.3, we review structural properties and particular cases of these problems. Works on heuristics are presented in Section 2.4 while polyhedral studies and exact algorithms are reviewed in Section 2.5.

We begin this chapter by introducing different concepts from graph theory and polyhedral theory that will be used later on. We assume that the reader is familiar with complexity theory, especially with the concept of NP-completeness. We refer to Garey and Johnson [39] for an in-depth treatment of these topics. For more information about graph theory concepts, we refer the reader to the book of Berge [8] or to Bondy and Murty [11], and to Nemhauser and Wolsey [71] for polyhedral theory and combinatorial optimization. Classical network flow problems and algorithms are studied in Ahuja et al. [4].

2.1 NOTATION AND DEFINITIONS

The aim in long-term planning of the backbone network is to determine a set of cables connecting given nodes and satisfying some survivability criteria that we will describe later. The given set of nodes and possible cable connections can be represented by an undirected graph $G = (V, E)$ where V is the set of *nodes* and E is the set of *edges* that represent the possible pairs of nodes between which a direct transmission link (cable) can be placed. The graph G may have parallel edges but should not contain loops. Graphs without parallel edges and without loops are called *simple*. If there exists an edge $e := \{i, j\}$ between two nodes i and j, these two nodes are called *adjacent*, and e is *incident* to i and j. Throughout this work, $n := |V|$ and $m := |E|$ will denote the number of nodes and edges of G.

Given the graph $G = (V, E)$ and $W \subset V$, the edge set

$$\delta(W) := \{\{i, j\} \in E \mid i \in W,\ j \in V \backslash W\}$$

is called the *cut* induced by W. We write $\delta_G(W)$ to make clear — in case of possible ambiguities — with respect to which graph the cut induced by W is considered. The *degree* of a node v is the cardinality of $\delta(v)$. The set

$$E(W) := \{\{i, j\} \in E \mid i \in W,\ j \in W\}$$

is the set of edges having both end nodes in W. We denote by $G(W) = (W, E(W))$ the subgraph induced by edges having both end nodes in W. If $E(W)$ is empty, W is an *independent set*. G/W is the graph obtained from G by contracting the nodes in W to a new node w (retaining parallel edges). Given two subsets of nodes W_1 and W_2, $W_1 \cap W_2 = \phi$, the subset of edges having one endpoint in each subset is denoted by

$$[W_1 : W_2] := \{\{i, j\} \in E \mid i \in W_1,\ j \in W_2\}.$$

We denote by $V - z := V \backslash \{z\}$ and $E - e := E \backslash \{e\}$ the subsets obtained by removing one node or one edge from the set of nodes or edges. $G - z$ denotes the graph $(V - z, E \backslash \delta(\{z\}))$, i.e. the graph obtained by removing a node z and its incident edges from G. This is extended to a subset $Z \subset V$ of nodes by the notation $G - Z := (V \backslash Z, E \backslash (\delta(Z) \cup E(Z)))$.

Each edge $e := \{i, j\} \in E$, has a *fixed cost* $c_e := c_{ij}$ representing the cost of establishing the direct link connection, and a *length* $d_e := d_{ij} := d(i, j)$. It is assumed throughout this work that these edge lengths satisfy the *triangle inequality*, i.e.

$$d(i, j) + d(j, k) \geq d(i, k) \quad \text{for all } i, j, k \in V.$$

The cost of a network $N = (V, F)$ where $F \subseteq E$ is a subset of possible edges is denoted by $c(F) := \sum_{e \in F} c_e$. The *distance* between two nodes i and j in this network is denoted by $d_F(i, j)$ and is given by the length of a shortest path linking these two nodes in F.

Without loss of generality, all costs are assumed to be nonnegative, because an edge e with a negative cost c_e will be contained in any optimum solution.

For any pair of distinct nodes $s, t \in V$, an *$[s, t]$-path* P is a sequence of nodes and edges $(v_0, e_1, v_1, e_2, \ldots, v_{l-1}, e_l, v_l)$, where each edge e_i is incident to the nodes v_{i-1} and v_i $(i = 1, \ldots, l)$, where $v_0 = s$ and $v_l = t$, and where no node or edge appears more than once in P. A collection $P_1, P_2, \ldots, P_k$ of $[s, t]$-paths is called *edge-disjoint* if no edge appears in more than one path, and is called *node-disjoint* if no node (other than s and t) appears in more than one path. A *cycle* (containing s and t) is a set of two node-disjoint $[s, t]$-paths.

A *Hamiltonian cycle* is a cycle using each node of the network exactly once. The problem of determining if a graph contains a Hamiltonian cycle is NP-complete. The corresponding optimization problem — the *traveling salesman problem (TSP)* — has been well studied. We refer to Lawler et al. [63] for an in depth treatment of this problem.

A graph $G = (V, E)$ is *k-edge-connected* (resp., *k-node-connected*) if, for each pair s, t of distinct nodes, G contains at least k edge-disjoint (resp., node-disjoint) $[s, t]$-paths. When the type of connectivity is not mentioned, we assume node-connectivity. The *edge connectivity* (resp., *node-connectivity*) of a graph is the maximal k for which it is k-edge-connected (resp., k-node-connected). A 1-edge-connected network is also 1-node-connected, and we call it simply *connected.* A cycle-free graph is a *forest* and a connected forest is a *tree*. A *connected component* of a graph is a maximal connected subgraph. If $G - e$ has more connected components than G for some edge e, we call e a *bridge*. Similarly, if

Z is a node set and $G - Z$ has more connected components than G, we call Z an *articulation set* of G. If a single node forms an articulation set, the node is called *articulation point*.

Node and edge-disjoint $[s,t]$-paths are related to cuts and articulation sets by Menger's theorem [67].

THEOREM 2.1 (MENGER)

1. *In a graph $G = (V, E)$, there is no cut of size $k - 1$ or less disconnecting two given nodes s and t, if and only if there exist at least k edge-disjoint $[s,t]$-paths in G.*

2. *Let s and t be two nonadjacent nodes in G. Then there is no articulation set Z of size $k - 1$ or less disconnecting s and t, if and only if there exist at least k node-disjoint $[s,t]$-paths in G.*

We will also use the following definitions arising from polyhedral theory. Given a polyhedron P, the *dimension* $\dim(P)$ of P is defined as the maximum number of affinely independent elements in P minus one. An inequality $a^T x \leq \alpha$ is *valid* with respect to P if $P \subseteq \{x : a^T x \leq \alpha\}$. The set $F_a := \{x \in P : a^T x = \alpha\}$ is called the *face* of P defined by $a^T x \leq \alpha$. If $dim(F_a) = \dim(P) - 1$ and $F_a \neq \phi$, then F_a is a *facet* of P and $a^T x \leq \alpha$ is called *facet-inducing* or *facet-defining*. A vector x is a *vertex* of P if it cannot be written as a non-trivial convex combination of points in P.

The convex hull of a set of points S will be denoted by $\text{conv}(S)$. We also denote by e_i the i-th unit vector in $\boldsymbol{R}^n$.

2.2 LOW-CONNECTIVITY CONSTRAINED NETWORK DESIGN PROBLEMS

Throughout this work, a (backbone) telephone network is seen as a set of gateway nodes (or telephone offices) and fiber links that are placed between nodes. In this context, survivability refers to the restoration of services in the event of office or link failure, or, in other words, a network is survivable if there exists a prespecified number of node-disjoint or edge-disjoint paths between any two offices. The only costs considered are construction costs, like the cost of digging trenches and placing a fiber cable into service.

In this framework, a considerable amount of research has focused on low-connectivity constrained network design problems. Following the terminology used by Monma and Shallcross [69] and Stoer [75], these models can be described informally as follows : we are given a set of telephone offices that have to be connected by a network. The offices may be classified according to importance, namely the

- *special offices*, for which a "high" degree of survivability has to be ensured in the network to be constructed;
- *ordinary offices*, which have to be simply connected to the network;
- *optional offices*, which may not be part of the network at all.

Given are also the pairs of offices between which a direct transmission link can be placed, and the associated cost of placing the fiber cable and putting it into service. The problem now consists in determining where to place fiber cables so that the construction cost, i.e. the sum of the fiber cable costs, is minimized and certain survivability constraints are ensured. For instance, we may require that

- the destruction of any single link may not disconnect any two special offices, or
- the destruction of any single office may not disconnect any two special offices.

These requirements are equivalent to ask that there exist

- at least two edge-disjoint paths, or
- at least two node-disjoint paths

between any two special offices.

Higher survivability levels may be imposed by requiring the existence of three or more paths between certain pairs of offices according to their importance class. However, up to now, low-connectivity requirements have been found to provide a sufficient level of survivability for telephone companies. For high-connectivity requirements, the reader is referred to Grötschel et al. [50, 52] and Stoer [75].

In graph-theoretic language, the set of offices and possible link connections can be represented by an undirected graph $G = (V, E)$. The survivability requirement or importance of a node is modeled by node types. In particular, each node $s \in V$ has an associated nonnegative integer r_s, the *type* of s. Sometimes, we also write $r(s)$ instead of r_s. A network $N = (V, F)$, where $F \subseteq E$ is a subset of the possible links, is said to satisfy the *node-connectivity requirements*, if, for each pair $s, t \in V$ of distinct nodes, N contains at least

$$r(s,t) := \min\{r_s, r_t\}$$

node-disjoint $[s, t]$-paths. Similarly, we say that N satisfies the *edge-connectivity requirements*, if, for each pair $s, t \in V$ of distinct nodes, N contains at least $r(s, t)$ edge-disjoint $[s, t]$-paths. If all node types have the same value k, it is equivalent to request that N is k-node-connected or k-edge-connected.

We restrict here to low-connectivity requirements, i.e. node types $r_s \in \{0, 1, 2\}$. Using our previous classification of nodes,

- special offices are represented by nodes of type 2,
- ordinary offices by nodes of type 1, and
- optional offices by nodes of type 0.

To shorten some notation, we extend the type function r to sets by setting

$$\begin{aligned} r(W) &:= \max\{r_s \mid s \in W\} \text{ for all } W \subseteq V, \text{ and} \\ \operatorname{con}(W) &:= \max\{r(s,t) \mid s \in W, t \in V \backslash W\} \\ &= \min\{r(W), r(V \backslash W)\} \\ &\qquad\qquad \text{for all } W \subseteq V, \phi \neq W \neq V. \end{aligned}$$

We write $\operatorname{con}_G(W)$ to make clear with respect to which graph $\operatorname{con}(W)$ is considered.

In order to formulate network design problems as integer linear programs, we associate with every subset $F \subseteq E$ an *incidence vector* $\mathbf{x} = (x_e)_{e \in E} \in \{0,1\}^{|E|}$ by setting

$$x_e := \begin{cases} 1 & \text{if } e \in F, \\ 0 & \text{otherwise}. \end{cases}$$

Conversely, each vector $\mathbf{x} \in \{0,1\}^{|E|}$ induces a subset

$$F^{\mathbf{x}} := \{e \in E \mid x_e = 1\}$$

of the edge set E. For any subset of edges $F \subseteq E$ we define

$$x(F) := \sum_{e \in F} x_e.$$

We can now formulate the connectivity constrained network design problem as the following integer linear program.

$$\min \sum_{e \in E} c_e x_e$$

s.t.

$$\begin{aligned}
& x(\delta(W)) \geq \text{con}(W) && W \subset V,\ \phi \neq W \neq V, && (2.2) \\
& x(\delta_{G-z}(W)) \geq \text{con}_{G-z}(W) && z \in V,\ W \subset V \backslash \{z\}, && \\
& && \phi \neq W \neq V \backslash \{z\}, && (2.3) \\
& x_e \in \{0,1\} && e \in E. && (2.4)
\end{aligned}$$

It follows from Menger's Theorem that, for any feasible solution $\mathbf{x}$ of this program, the subgraph $N = (V, F^{\mathbf{x}})$ of G defines a network satisfying the node-connectivity requirements. Removing (2.2), we obtain an integer linear program for edge-connectivity requirements. Inequalities (2.1) are called *cut inequalities*, while inequalities (2.2) are called *node cut inequalities*.

In the next sections, we give a review of the work on these models, describing heuristics and exact solution methods for more general or more specialized problems. Much of the material is taken from references cited in the surveys of Christofides and Whitlock [19], Winter [87] and Stoer [75].

2.3 STRUCTURAL PROPERTIES AND PARTICULAR CASES

A lot of research has focused on the survivability model presented in the previous section. The next sections survey these results. We begin by looking at the complexity of the problem, before considering some polynomially solvable cases. We also present work on particular cases, either with restricted connectivity requirements or restricted costs.

Complexity

The connectivity constrained network design problem is NP-hard in general. In particular :

- If $r_s \in \{0,1\}$, $\forall s \in V$, it reduces to the well-known NP-hard Steiner tree problem in networks. Winter [88] made an in depth survey of these problems.

- If $r_s = 2$, $\forall s \in V$, it consists in determining a minimum cost two-connected network. This last problem is NP-hard even if the graph is complete and costs satisfy the triangle inequality, since with an algorithm for this problem, one could decide whether a graph has a Hamiltonian cycle by associating a cost equal to 1 to all graph edges and cost equal to 2 to all non-graph edges (see Eswaran and Tarjan [30]).

However, for some particular connectivity requirements or costs, or when the underlying graph G is restricted, the problem may become polynomially solvable. We now review these cases.

Restricted connectivity requirements

By restricting the connectivity requirements r_s, the connectivity constrained network design problem reduces to some well-known polynomially solvable problems :

- If $r_s = 1$, $\forall s \in V$, the problem reduces to the minimum spanning tree problem. The most famous polynomial time algorithms for solving it are those from Kruskal [61] and Prim [72].
- If $r_s = 1$ for exactly two nodes of V and $r_s = 0$ for all the other nodes, the problem becomes a shortest path problem, solvable e.g. by the algorithms of Bellman [7] or Dijkstra [29].
- If $r_s = k$, $k \geq 2$, for exactly two nodes of V and $r_s = 0$ for all the other nodes, the problem becomes a k-shortest paths problem. This problem was studied by Suurballe [76] and Suurballe and Tarjan [77].
- If $r_s \in \{0, 1\}$, $\forall s \in V$, the problem reduces to the Steiner tree problem in networks. This problem is NP-hard in general, but Lawler [62] solved it in the case where either the number of nodes of type 0 or the number of nodes of type 1 is restricted.

Restricted costs

Under uniform or 0/1 costs, certain classes of connectivity constrained network design problems are polynomially solvable. We now examine these choices of costs.

Under uniform costs, the underlying graph G can be seen as a complete graph, and the problem turns into the construction of a sufficiently highly connected graph with a minimum number of edges. Chou and Frank [17] solved this problem for edge connectivity requirements by producing a graph where each node has degree r_s, except possibly for one node that has degree $r_s + 1$. Since these are the lowest possible degrees under the given connectivity requirements, the graph has the minimum number of edges. This proves the following lemma.

LEMMA 2.2 *Given node types $r_s \geq 2$ for a set V of nodes, the minimum number of edges of a graph satisfying the edge-connectivity requirements given by r is*

$$\left\lceil \frac{1}{2} \sum_{s \in V} r_s \right\rceil .$$

The use of parallel edges is allowed in the construction.

Stoer [75] describes a polynomial algorithm similar to that of Chou and Frank which also handles nodes of type 1. Frank and Chou [34] also solved the problem when no parallel edges but extra nodes are allowed in the solution.

Unfortunately, to our knowledge, no general solution for the node connectivity version of the problem is available in the literature. But more can be said about uniform connectivity requirements $r_s = k$ for some $k \geq 2$.

An early work by Fulkerson and Shapley [37] — written in 1961 but published ten years later — proved Lemma 2.2 for the edge-connectivity problem with uniform requirements, but without using parallel edges. Harary [53] showed with the help of a polynomial algorithm that the same result holds for the node-connectivity problem with uniform requirements, leading to the following lemma.

LEMMA 2.3 *Given $k \geq 2$ and $n \geq k+1$, the minimum number of edges in a k-node-connected graph on n nodes without parallel edges is*

$$\left\lceil \frac{kn}{2} \right\rceil .$$

Now, one may guess that Lemma 2.2 also holds for general node-connectivity requirements, but this conjecture is not true and a counter-example can be found in Stoer [75].

We now turn to problems with 0/1 costs. These are known in the literature as augmentation problems, since these correspond to the problem of augmenting a graph $G = (V, E)$ by a minimum number of edges in $V \times V$, so that it meets connectivity requirements.

These augmentation problems were brought up by Eswaran and Tarjan [30] for two-edge and two-node-connected graphs. Rosenthal and Goldner [73] studied the augmentation to two-node-connected graphs. Their linear time algorithm contains an error that was corrected by Hsu and Ramachandran [56], who also proposed a parallel implementation of their algorithm. Hsu and Ramachandran [55] also developed a linear time algorithm for the augmentation to 3-node-connected networks. The augmentation to k-edge-connected graphs was studied by Watanabe and Nakamura [82], Ueno et al. [80] and Cai and Sun [14]. The fastest known algorithm for this problem is the one by Naor et al. [70]. Frank [33] solved the augmentation problem completely for general edge-connectivity requirements. All solution procedures allow the use of parallel edges, except those of Eswaran and Tarjan [30] and Rosenthal and Goldner [73]. Again, the problem of augmentation to a node-connected graph is open in most cases.

Other polynomially solvable cases

Other cases of connectivity constrained network design problems are polynomially solvable if the underlying graph G is restricted to certain graph classes.

Among these, the class of *series-parallel graphs* has received a lot of attention. Series-parallel graphs are created from a single edge by two operations :

- addition of parallel edges,
- subdivision of edges by insertion of nodes.

Works on various connectivity requirements for these graphs can be found in the literature :

- $r_s \in \{0, 1\}$, $\forall s \in V$ (Steiner tree problem) :
  The problem was solved in linear time by Wald and Colbourn [81]. Goemans [42] gave a complete description of the polytope associated with the solutions of the problem.

- $r_s = k$, $k \geq 2$, $\forall s \in V$, with edge connectivity requirements (k-edge-connected network problem) :
  Mahjoub [66] gave a complete description of the polytope associated with the solutions of the case when $k = 2$. This work was extended to any $k \geq 2$ by Didi Biha and Mahjoub [28].

- $r_s \in \{0, 2\}$, $\forall s \in V$ (two-connected Steiner subgraph problem) :
  The problem was solved in linear time by Winter [86], both for edge and node-connectivity requirements. Coullard et al. [23] gave a complete description of the polytope associated with the solutions of the node-connectivity case.

Winter has also developed linear-time algorithms for the case $r_s \in \{0, 2\}$ in outerplanar [85] and Halin graphs [84]. He also mentions in [88] that he solved the problem in linear time for $r_s \in \{0, 3\}$ in Halin graphs. Coullard et al. solved the problem with $r_s \in \{0, 2\}$ in W_4-free graphs [25] and gave a complete description of the dominant of the corresponding polytope [24].

Dominant of the polytopes of k-edge-connected networks where parallel edges are allowed were completely described by Cornuéjols et al. [22] for k even and G series-parallel and by Chopra [16] for k odd and G outerplanar.

Structural results

As we already mentioned, the problem of determining a minimum cost two-connected network ($r_s = 2$, $\forall s \in V$) is NP-hard even if the graph is complete and the costs satisfy the triangle inequality. In this case, however, it is possible to compute lower bounds and find worst-case guarantees for some heuristics. We review these heuristics in the next section.

Demanding two-edge-connectivity in a complete graph with costs satisfying the triangle inequality is actually the same as demanding two-node connectivity, because a minimum cost two-edge-connected graph is either already two-node-connected or can be made two-node-connected without increase of cost, as shown by Frederickson and Já Já [36].

Monma et al. [68] obtained a structural characterization of optimal two-connected solutions with costs satisfying the triangle inequality : they showed that there exists an optimal solution whose nodes all have degree 2 or 3 and such that the removal of any edge or pair of edges leaves a bridge in the resulted connected components. This work was extended to k-connected networks by Bienstock et al. [9], showing that an optimal k-edge connected graph satisfies the following conditions :

1. every node has degree k or $k+1$;

2. removing any $1, 2, \ldots$ or k edges does not leave the resulting components all k-edge-connected.

They also proved that a minimum cost k-node-connected graph does not necessarily have the same cost as a minimum cost k-edge connected graph. However, the characterization of optimal solutions holds for k-node-connected graphs (by replacing edge by node) if $|V| \geq 2k$.

Another interesting result, under the triangle inequality assumption, was proved by Monma et al. [68]. They showed that the cost of an optimal Hamiltonian cycle is at most $\frac{4}{3}$ times the cost of an optimal two-connected graph. This was also proved by Frederickson and Já Já [36] in the particular case where the minimum cost two-connected graph is planar.

2.4 HEURISTICS

In this section, we present heuristics found in the literature for solving certain classes of connectivity constrained network design problems. Many heuristics and exact algorithms are known for the Steiner tree problem. Since we are more interested here in higher-connectivity requirements, we refer the reader to the surveys of Winter [88] and Maculan [64] on the Steiner tree problem.

Constructive and improvement heuristics

The first heuristics for designing minimum cost survivable networks appear in the early work of Steiglitz et al. [74]. They consider the problem under general node-connectivity requirements r. Their approach is to find a starting solution in a randomized greedy way, testing for feasibility at each step. Once a feasible solution is found, an optimizing routine tries to improve this solution by applying a transformation similar to the 2-opt exchange for the TSP, until no more improvements can be found.

Monma and Shallcross [69] considered the low-connectivity problems with node types in $\{0, 1, 2\}$. They developed several heuristics for constructing initial feasible networks, and local improvement heuristics.

Their constructive heuristics exploit the structure of two-edge or two-node connected networks. Any two-connected network can be constructed by an *ear-composition* procedure. That is, a first partial solution is formed by constructing a cycle on a subset of the nodes. Then, paths are repeatedly added to the solution until all nodes are included. These paths — called *ears* — start at one node on the partial solution, pass through nodes not yet visited, and end back at a different node on the solution. Monma and Shallcross developed two particular ear-inserting strategies : a greedy procedure that tries to produce the best solution possible, and a random one to construct a variety of starting solutions. In Section 7.1, we describe how we adapted the ear-inserting procedure to solve the 2CNBR problem.

Monma and Shallcross also proposed the two-trees heuristic, that constructs a two-connected network by augmenting a spanning tree T by another tree spanning the leaves of T.

Their improvement heuristics are based on local exchange techniques. These techniques are :

1-opt : replaces one edge with another while maintaining feasibility.

2-opt and 3-opt : similar to the k-opt exchanges for the TSP, these replace two (resp. three) edges lying on one cycle by two (resp. three) edges in order to construct a different cycle.

pretzel : replaces one edge by two crossing edges to break a cycle in two cycles.

quetzel : reverse of pretzel; merges two cycles into one.

degree : based on the degree result of Monma et al. [68] for costs satisfying the triangle inequality, this procedure reduces the degree of nodes.

These low-connectivity heuristics were modified by Ko and Monma [60] for the design of k-edge or k-node-connected graphs.

Clarke and Anandalingam [20] add a constructive heuristic to those proposed by Monma and Shallcross for the two-edge-connected network problem. The *bootstrap heuristic* uses the solution of a simple linear programming relaxation of the problem and transforms it into a feasible solution by rounding to one all variables with a positive value. This way of generating feasible solutions is classical in a Branch-and-Cut framework, and we used it in conjunction with our stingy heuristic — described in Section 7.4 — to get good upper bounds in our Branch-and-Cut algorithm for solving the 2CNBR problem.

Clarke and Anandalingam improve the solutions obtained by the bootstrap heuristic by applying 1-opt, 2-opt and pretzel exchanges, as well as a *delete step* that removes edges in a way analog to our stingy heuristic.

Both the heuristics of Monma and Shallcross and of Clarke and Anandalingam are part of packages for the design of survivable networks with additional constraints concerning routing of the expected demands. These packages are described in Cardwell et al. [15] and Clarke and Anandalingam [21]. De Jongh [27] made a survey of models for survivable network design that include routing aspects.

Heuristics with worst-case guarantee

As we already mentioned, if costs satisfy the triangle inequality, the cost of an optimal Hamiltonian cycle is at most $\frac{4}{3}$ times the cost of an optimal two-connected graph (Monma et al. [68]). Using Christofides heuristic [18], having a worst-case guarantee of $\frac{3}{2}$ for the TSP, we obtain a polynomial time heuristic for the two-connected network problem with a worst-case guarantee of 2. More precisely, if H denotes the solution of Christofides heuristic, T the optimal Hamiltonian tour and TC the optimal two-connected network, we have

$$c(H) \leq \frac{3}{2}\, c(T) \leq \frac{3}{2}\left(\frac{4}{3}\, c(TC)\right) = 2\, c(TC).$$

Frederickson and Já Já [36] modified Christofides heuristic for the design of two-connected networks and achieved worst-case of $\frac{3}{2}$.

The same authors [35] also provided an approximation algorithm for the problem of increasing the edge-connectivity of any graph to 2 with general costs. Their algorithm runs in $O(n^2)$ time and has a worst-case of 2 if the starting graph is connected, 3 otherwise.

Khuller and Thurimella [58] improved this result by developing an algorithm with a worst-case of 2, running in $O(m + n \log n)$ time, for the augmentation to two-node or two-edge-connectivity, starting with a connected graph. They also provided an algorithm running in $O(nk \log n(m + n \log n))$ that increases the edge-connectivity of a graph (not necessarily connected) to k with a worst case of 2. Khuller and Vishkin [59] developed $O(kn^3 \log n)$ running time algorithms for two-edge and two-node-connected network problems with unit costs. These algorithms have worst-cases of respectively $\frac{3}{2}$ and $\frac{5}{3}$. Khuller [57] made a survey of these approximation algorithms for uniform edge and node connectivity requirements.

For general edge-connectivity requirements r, Goemans and Bertsimas [43], Agrawal et al. [2] and Goemans and Williamson [44] described approximation algorithms in which edges can be selected several times. Their worst-case ratio is

$$2\left(\sum_{k=1}^{l} \frac{\rho_k - \rho_{k-1}}{\rho_k}\right)$$

where $\rho_l > \rho_{l-1} > \ldots > \rho_0 = 0$ are the different values of the connectivity requirements. In particular, for two-edge connected networks, this leads to a worst-case ration of 2.

More recently, Williamson et al.[83] studied the case where no parallel edges are allowed and achieved a worst-case of

$$2\sum_{k=1}^{l} \mathcal{H}(\rho_k - \rho_{k-1})$$

where $\mathcal{H}$ is the harmonic function $\mathcal{H}(k) = 1 + \frac{1}{2} + \frac{1}{3} + \ldots + \frac{1}{k}$. This worst-case is equal to 3 for two-edge connected networks. If $\rho_1 = 1$ and $l \geq 2$, then the performance guarantee improves by one unit. Gabow et al. [38] improved the running time of this algorithm from $O(k^3n^4)$ to $O(kn^2(k + \sqrt{\log\log n}))$. These algorithms are based on the primal-dual approximation algorithm that was invented by Goemans and Williamson. In [45], they present this method and apply it to a wide variety of NP-hard network design problems.

2.5 POLYHEDRAL STUDIES AND EXACT ALGORITHMS

In Section 2.3, we mention certain classes of connectivity constrained network design problems for which the corresponding polytope has been completely described. We now turn to polyhedral studies and Branch-and-Cut algorithms for the general case. Most of these works are based on the linear programming formulation (2.1)-(2.3) given in Section 2.2.

The first method for solving the problem exactly was developed by Christofides and Whitlock [19]. Their algorithm is based on the linear relaxation obtained by replacing integrality constraints (2.3) by $0 \leq x_e \leq 1$, and only keeping cut constraints (2.1) corresponding to subsets W such that $|W| = 1$. These particular cut constraints are called *degree constraints*. The starting linear program is thus

$$\min \sum_{e \in E} c_e x_e$$

s.t.

$$x(\delta(v)) \geq r(v) \qquad v \in V, \tag{2.5}$$

$$0 \leq x_e \leq 1 \qquad e \in E. \tag{2.6}$$

Given a solution to this LP, they impose the edge-connectivity requirements by adding violated cut constraints. These can be found in polynomial time by computing the minimum cut in the graph $G = (V, E)$ with edge capacities equal to the values of the corresponding variables in the solution of the current LP — using e.g. the Gomory-Hu algorithm [46]. When all cut constraints are satisfied, if some variables have fractional values, a branch-and-boundprocedure is applied. Christofides and Whitlock mention in [19] that this algorithm is able to solve problems with "well over a hundred nodes" for edge-connectivity requirements.

If one wants to add node-connectivity requirements, they propose to check the node-connectivity each time an integer solution satisfying edge-connectivity requirements is found. If some node-connectivity requirements are violated, the corresponding node-cut constraints (2.2) are added to the LP.

Grötschel, Monma and Stoer studied in detail network design problems with connectivity constraints. A survey of their work can be found in [51] and [75].

In their earliest work on the subject, Grötschel and Monma [47] introduced a general model mixing edge and node survivability requirements. They examined the dimension of the associated polytope and proved facet results for cut and node-cut inequalities.

They also described completely the polytope of the (1-)connected network problem, based on the work of Cornuéjols et al. [22]. This is done by the introduction of *partition inequalities*, that generalize cut inequalities (2.1). Given a partition $W_1, W_2, \ldots, W_p$ ($p \geq 2$) of V into p nonempty subsets, the inequality

$$\frac{1}{2}\sum_{i=1}^{p} x(\delta(W_i)) \geq p-1$$

is valid for the polytope of connected networks.

Based on partition inequalities for connected networks, Grötschel and Monma introduced the *node-partition inequalities* for k-node-connected networks. These inequalities come from the fact that the deletion of $k-1$ nodes from a k-node-connected network leaves a connected graph. Thus, if $Z \subseteq V$ is a node set with exactly $k-1$ nodes and $W_1, W_2, \ldots, W_p$ ($p \geq 2$) is a partition of $V \backslash Z$ into p nonempty subsets, the inequality

$$\frac{1}{2}\sum_{i=1}^{p} x(\delta_{G-Z}(W_i)) \geq p-1$$

is valid for the polytope of k-node-connected networks. It is obvious that these node-partition inequalities are a generalization of node cut inequalities (2.2) .

Grötschel, Monma and Stoer then attacked low-connectivity constrained problems (with $r_s \in \{0,1,2\}$), deriving new facets [49] and implementing some of these into a Branch-and-Cut algorithm [48]. They generalized partition and node-partition inequalities, and introduced lifted 2-cover and comb inequalities. These results were extended to higher-survivability requirements in [52].

More inequalities for two-edge-connected network problemswere found by Boyd and Hao [12] (complemented comb inequalities), and by Boyd and Zhang [13] (clique tree inequalities).

Chapter 3

TWO-CONNECTED NETWORKS WITH BOUNDED RINGS: THE MODEL

This chapter is devoted to the description of our model for designing survivable networks. The first section explains what reasons led us to develop this new model, and gives a description of the problem in terms of connectivity and cycle lengths. Mathematical formulations of the *Two-Connected Network with Bounded Rings problem (2CNBR)* are given in the second section.

3.1 MOTIVATION

When Belgacom, the Belgian telephone operator, asked us to develop tools for long-term planning in the backbone network, we first looked at the existing models described in the previous chapter.

Long-term planning involves a ten-year timescale. Demand data for such a long period are not reliable enough, and we only deal with topological aspects. In this case, we want to determine a set of cables connecting all nodes, according to some survivability criteria. Since we consider only construction costs, survivability models based on connectivity requirements seemed well suited. Furthermore, two-node-connectivity was a sufficient level of survivability for the Belgian backbone network, since the probability of dealing with two simultaneous failures is very low.

We thus wanted to solve the problem of designing a minimum cost two-connected network, which can be formulated as the following integer linear program :

$$\min \sum_{e\in E} c_e x_e$$

s.t.

$$x(\delta(W)) \geq 2 \qquad W \subset V,\ \phi \neq W \neq V, \tag{3.2}$$

$$x(\delta_{G-z}(W)) \geq 1 \qquad z \in V,\ W \subset V\backslash\{z\},\ \phi \neq W \neq V\backslash\{z\}, \tag{3.3}$$

$$x_e \in \{0,1\} \qquad e \in E. \tag{3.4}$$

An easily implementable method to solve it exactly is the Branch-and-Cut algorithm proposed by Christofides and Whitlock [19], which is described in Section 2.5. We managed to solve the problem on the 52 nodes of the Belgian backbone network using only cut constraints (3.1) and a few branching steps. The optimal solution of the problem is the Hamiltonian cycle depicted in Figure 3.1. Since this solution is a tour, any edge failure would require to reroute the flow that passed through that edge, using all the edges of the network, an obviously undesirable feature.

Our aim was then to add extra constraints to limit the region of influence of the traffic which is necessary to reroute if a connection is broken. Imposing a limit on the length of the rerouting can be done by limiting the length of the shortest cycle including each edge. Such a condition has also a direct implication in topologies using the emerging technology of *self-healing rings*. Self-healing rings are cycles in the network equipped in such a way that any link failure in the ring is automatically detected and the traffic rerouted by the alternative path in the cycle. It is natural to impose a limited length of these rings. This is equivalent to set a bound on the length of the shortest cycle including each edge.

The *Two-Connected Network with Bounded Rings problem (2CNBR)* is the problem of designing a minimum cost network N satisfying the following constraints:

1. N contains at least two node-disjoint paths between every pair of nodes (*two-connectivity constraints*),

2. each edge of N must belong to at least one cycle whose length is bounded by a given constant K (*ring constraints*).

We saw in Chapter 1 that two-connectivity constraints have been thoroughly studied. However, to our knowledge, ring constraints were not considered before.

3.2 MATHEMATICAL FORMULATIONS

In this section, we present two formulations of the 2CNBR problem as integer linear program. Both formulations consider two-connectivity and ring constraints separately. Two-connectivity is obtained by cut and node-cut constraints (3.1)-(3.2). To impose ring constraints, we use extra variables. The first formulation uses an explicit enumeration of all feasible cycles, associating a new variable to each cycle. The ring constraints are then expressed using covering constraints. In the second formulation, cycles are expressed using flow variables.

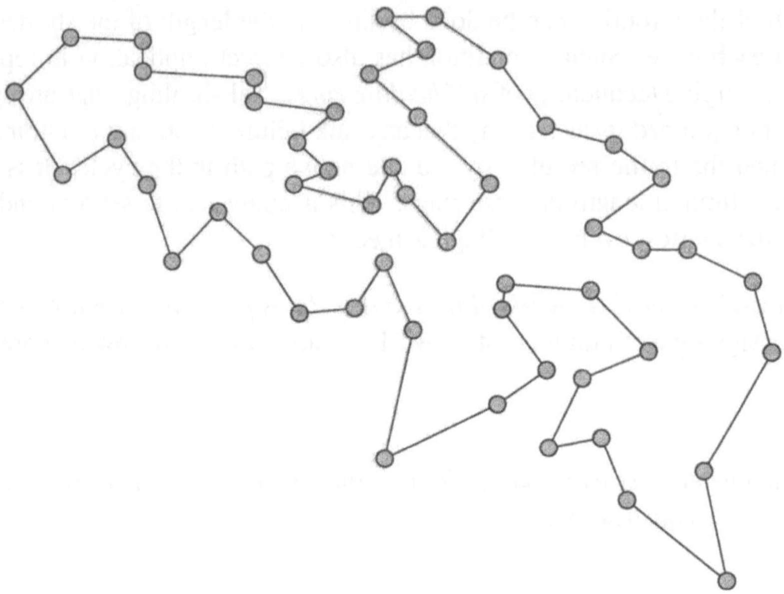

Figure 3.1. Minimum cost two-connected network

Complete enumeration of feasible rings

The ring constraints can be formulated as a covering problem, since bounded rings must cover the edges of the network. This requires a variable associated to each cycle of bounded size. This formulation was introduced by Fortz et al. [31, 32]. Here, we present a strengthened form of it.

To do so, we introduce the following notation :

- $\mathcal{C}_e$, $e \in E$, is the set of cycles in $G = (V, E)$ that include edge e and whose length is less than or equal to K;

- y_e^c, $c \in \mathcal{C}_e$, $e \in E$, is a binary variable such that

$$y_e^c = \begin{cases} 1 & \text{if all edges in cycle } c \\ & \qquad \text{are included in the solution } F; \\ 0 & \text{otherwise.} \end{cases}$$

It is then sufficient to impose that

1. each edge in the solution belongs to one feasible cycle;
2. a cycle is present in the solution only if its edges are present.

In other words, this second condition is equivalent to

$$y_e^c = 1 \qquad \text{iff} \qquad \forall f \in c,\ x_f = 1. \tag{3.5}$$

The fact that the cycles must cover the edges that are in the solution together with condition (3.4) can be expressed as the following system of inequalities :

$$y_e^c \le x_f \qquad e \in E,\ c \in \mathcal{C}_e,\ f \in c, \tag{3.6}$$

$$\sum_{c \in \mathcal{C}_e} y_e^c \ge x_e \qquad e \in E. \tag{3.7}$$

It is possible to strengthen condition (3.5) in the following way : given a feasible solution F and an edge $e \in F$, suppose that two different feasible cycles, $c, c' \in \mathcal{C}_e$, are in F. Then we can have together $y_e^c = 1$ and $y_e^{c'} = 1$. But it is obvious that we only need one of the two variables to be equal to one to satisfy condition (3.6). It means that for each edge $f \in c \cap c'$, we can write

$$y_e^c + y_e^{c'} \le x_f,$$

which is stronger than (3.5). More generally, we can replace (3.5) by

$$\sum_{c \in \mathcal{C}_e:\ f \in c} y_e^c \le x_f \qquad e \in E,\ f \in E \backslash \{e\}. \tag{3.8}$$

The introduction of these stronger inequalities also reduces the number of constraints needed to formulate the 2CNBR problem.

A complete formulation of the 2CNBR problem is thus given by :

$$\min \sum_{e \in E} c_e x_e$$

s.t.

$$x(\delta(W)) \geq 2 \qquad W \subset V,\ \phi \neq W \neq V, \tag{3.9}$$

$$x(\delta_{G-z}(W)) \geq 1 \qquad z \in V,\ W \subset V\backslash\{z\},\ \phi \neq W \neq V\backslash\{z\}, \tag{3.10}$$

$$\sum_{c \in \mathcal{C}_e:\ f \in c} y_e^c \leq x_f \qquad e \in E,\ f \in E\backslash\{e\}, \tag{3.11}$$

$$\sum_{c \in \mathcal{C}_e} y_e^c \geq x_e \qquad e \in E, \tag{3.12}$$

$$x_e \in \{0,1\} \qquad e \in E, \tag{3.13}$$

$$y_e^c \in \{0,1\} \qquad e \in E, c \in \mathcal{C}_e. \tag{3.14}$$

This formulation involves

- $m + Q$ variables, where $Q := \sum_{e \in E} |\mathcal{C}_e|$ is the total number of possible cycles,
- $O(n2^n)$ cut and node-cut inequalities ,
- $O(m^2)$ covering constraints.

Since a large number of variables and constraints appear here, this formulation cannot be used directly. The difficulty arising from the exponential number of cut and node-cut constraints can be avoided by using a cutting plane approach, since these can be separated in polynomial time, as we already mentioned. But the number of y-variables and the number of covering constraints is also too large. For example, if G is a complete graph and all edges have unit length, it is easy to see that there are $O(n^K)$ y-variables and $O(n^4)$ set covering constraints. If $n = 50$ and $K = 4$, we have 249,900 y-variables and 1,501,850 set covering constraints. A cutting plane approach enables us to deal with the large number of constraints.

Furthermore, by projecting the set covering formulation on the x-space, we obtain valid inequalities allowing us to restrict the linear program to the m design variables. This approach is described in Section 4.5.

Flow formulation

Another formulation is obtained using flow variables to represent cycles. In Section 2.1, we defined a cycle using nodes i and j as a pair of node-disjoint $[i, j]$-paths. If an edge $e := \{i, j\}$ belongs to the solution of 2CNBR, two node-disjoint $[i, j]$-paths of total length less than or equal to K must exist. Since one of these paths is e itself — by the triangle inequality —, node-disjoint is here equivalent to edge-disjoint, and two edge-disjoint $[i, j]$-paths can be seen as a flow of value 2 between i and j.

For each edge $\{i, j\} \in E$, we introduce $2m$ new variables $u_{kl}^{\{i,j\}}$, $k \neq l$, $\{k, l\} \in E$ that represent the (directed) flow going from k to l, corresponding to the feasible cycle for edge $\{i, j\}$.

The following conditions must hold for these new variables :

1. $u^{\{i,j\}}$-variables represent a flow of value 2 if $\{i, j\}$ is included in the solution, 0 otherwise;

2. the capacity from k to l is equal to 1 if the edge $\{k, l\}$ is in the solution, 0 otherwise;

3. the cost for sending one unit of flow from k to l is $d(k, l)$;

4. the cost of the flow represented by $u^{\{i,j\}}$-variables is less than or equal to K.

These conditions, together with cut and node-cut constraints, lead to the following integer linear program :

$$\min \sum_{e \in E} c_e x_e$$

s.t.

$$
\begin{aligned}
& x(\delta(W)) \geq 2 && W \subset V,\ \phi \neq W \neq V, && (3.15)\\
& x(\delta_{G-z}(W)) \geq 1 && z \in V,\ W \subset V \backslash \{z\}, & \\
& && \phi \neq W \neq V \backslash \{z\}, && (3.16)\\
& \sum_{l:\ \{k,l\} \in E} (u_{kl}^{\{i,j\}} - u_{lk}^{\{i,j\}}) = \begin{cases} 2\, x_{\{i,j\}} & \text{if } k = i \\ -2\, x_{\{i,j\}} & \text{if } k = j \\ 0 & \text{otherwise} \end{cases} && & \\
& && \{i,j\} \in E,\ k \in V, && (3.17)\\
& u_{kl}^{\{i,j\}} + u_{lk}^{\{i,j\}} \leq x_{\{k,l\}} && \{i,j\} \in E,\ \{k,l\} \in E, && (3.18)\\
& \sum_{\{k,l\} \in E} d(k,l)(u_{kl}^{\{i,j\}} + u_{lk}^{\{i,j\}}) \leq K, && & \\
& && \{i,j\} \in E, && (3.19)\\
& u_{kl}^{\{i,j\}} \geq 0 && \{i,j\} \in E,\ \{k,l\} \in E, && (3.20)\\
& u_{lk}^{\{i,j\}} \geq 0 && \{i,j\} \in E,\ \{k,l\} \in E, && (3.21)\\
& x_{\{i,j\}} \in \{0,1\} && \{i,j\} \in E. && (3.22)
\end{aligned}
$$

Equations (3.16) are flow conservation constraints that ensure that $u^{\{i,j\}}$-variables represent a flow of value 2 if $\{i,j\}$ is included in the solution. Inequalities (3.17) are capacity constraints and inequalities (3.18) impose the bound on the cost of the flows. Note that integrality of $u_{kl}^{\{i,j\}}$ is implied by the integrality of $x_{\{i,j\}}$.

This formulation involves

- $m + 2m^2$ variables,
- $O(n2^n)$ cut and node-cut inequalities,
- $O(m^2)$ inequalities arising from the flow formulation of the ring constraints.

Again, the number of variables and constraints is too large for being used in practice, but this formulation can be projected on the x-space, as we will see in Section 4.6.

Chapter 4

POLYHEDRAL STUDY

The dominant of the polytope associated to the feasible solutions of the 2CNBR problem is studied in this chapter. We define this polytope in Section 4.1 and examine its dimension. Trivial facets are also examined. Valid inequalities for the 2CNBR problem are studied in Sections 4.2 to 4.8. We also study conditions under which some classes of valid inequalities are facet-defining.

4.1 ASSOCIATED POLYTOPES AND TRIVIAL INEQUALITIES

Given a graph $G = (V, E)$ and a constant $K > 0$, we define

$$\begin{aligned} \mathcal{P}_{G,K} &:= \text{conv}\{x \in \boldsymbol{R}^{|E|} : \exists y \in \boldsymbol{R}^{Q},\ x, y \text{ satisfy (3.8)-(3.13)}\} \\ &= \text{conv}\{x \in \boldsymbol{R}^{|E|} : \exists u \in \boldsymbol{R}^{2|E\times E|}, \\ &\qquad x, u \text{ satisfy (3.14)-(3.21)}\} \end{aligned}$$

as the polytope associated to the 2CNBR problem. This polytope is the convex hull of the projection of the two formulations of the problem on the x-space.

Any feasible solution of 2CNBR is a vertex of this polytope. Since we supposed that all costs are nonnegative, the optimal solution of 2CNBR is also a vertex of the *dominant* of $\mathcal{P}_{G,K}$, defined as

$$\mathcal{D}om\mathcal{P}_{G,K} := \text{conv}\{x \in \boldsymbol{R}^{|E|} : \exists y \in \mathcal{P}_{G,K},\ x \geq y\}.$$

Note that for the models presented in Chapter 2, i.e. survivable network design problems without bounded ring constraints, the convex hull of 0/1-vectors in the dominant of the polytope of feasible solutions is the polytope itself. In particular, the polytope of two-connected networks is obtained from the 2CNBR problem by relaxing the bounded ring constraints, i.e. taking $K = \infty$, and we have $conv(\mathcal{D}om\mathcal{P}_{G,\infty} \cap \{0,1\}^{|E|}) = \mathcal{P}_{G,\infty}$. However, this is not true for the general case : consider, for example, the graph depicted in Figure 4.1 with unit edge lengths. Let $K = 4$ be the bound on the ring lengths. Then the incidence vector of the graph does not belong to $\mathcal{P}_{G,K}$ — since edge $\{10, 12\}$ does not belong to a feasible ring — while it belongs to $\mathcal{D}om\mathcal{P}_{G,K}$, as the graph contains a feasible solution.

To analyze the dominant of $\mathcal{P}_{G,K}$, we would like to characterize under which conditions the incidence vector of a graph belongs to $\mathcal{D}om\mathcal{P}_{G,K}$. A useful tool to derive these conditions is the *restriction of G to bounded rings* defined as follows.

DEFINITION 4.1 (RESTRICTION OF A GRAPH TO BOUNDED RINGS) *Let $G = (V, E)$ be a graph and $K > 0$ a given constant. For each subset of edges*

$F \subseteq E$, we define its restriction to bounded rings F_K as

$$F_K := \left\{ e \in F : \begin{array}{l} \text{e belongs to at least one cycle} \\ \text{of length less than or equal to } K \text{ in } F \end{array} \right\}.$$

The subgraph $G_K = (V, E_K)$ is the restriction of G to bounded rings.

REMARK 4.1 *An edge $e \in E \backslash E_K$ will never belong to a feasible solution of 2CNBR.*

REMARK 4.2 *The incidence vector of $F \subseteq E$ belongs to $\mathcal{D}om\mathcal{P}_{G,K}$ if and only if F_K is two-connected.*

REMARK 4.3 *$\mathcal{D}om\mathcal{P}_{G,K}$ is fully dimensional.*

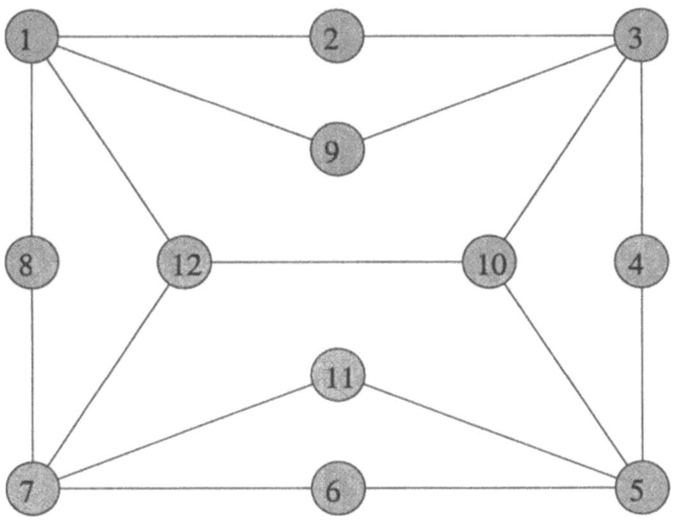

Figure 4.1. Example of graph belonging to $\mathcal{D}om\mathcal{P}_{G,K}$ but not to $\mathcal{P}_{G,K}$

Remark 4.1 shows that we can eliminate variables corresponding to edges in $E \backslash E_K$, and thus suppose that $G_K = G$. This allows us to derive sufficient conditions under which $\mathcal{P}_{G,K}$ is fully dimensional.

PROPOSITION 4.4 *Let $G = (V, E)$ be a graph, $K > 0$ a given constant and $G_K = (V, E_K)$ the restriction of G to bounded rings. If*

1. *$G = G_K$,*
2. *$G - e$ is a feasible solution of 2CNBR for all $e \in E$,*

then $\mathcal{P}_{G,K}$ is fully dimensional.

PROOF.

We will show that there are $m + 1$ affinely independent points in the polytope. It follows from Remark 4.2 and from the above conditions that the incidence vector of E and $E \backslash \{e\}$, for all $e \in E$, belong to $\mathcal{P}_{G,K}$. These $m + 1$ vectors are affinely independent vectors of $\mathcal{P}_{G,K}$, which is thus fully dimensional.

□

Almost all our facet results apply to the dominant $Dom\mathcal{P}_{G,K}$. Throughout the remaining of this work, we will assume that $G = G_K$. The following result characterizes which of the trivial inequalities $x_e \geq 0$ define facets of $Dom\mathcal{P}_{G,K}$.

THEOREM 4.5 *Let $G = (V, E)$ be a graph, $K > 0$ a given constant and $e \in E$ an edge of the graph.*

$x_e \geq 0$ defines a facet of $Dom\mathcal{P}_{G,K}$ if and only if for every edge $f \neq e$, $(G - \{e, f\})_K$ is two-connected.

PROOF.

If $(G - \{e, f\})_K$ is not two-connected for some $f \in E$, $f \neq e$, then $x_e = 0$ implies $x_f = 1$ for all $x \in Dom\mathcal{P}_{G,K}$ but not reversely. Therefore, $x_e + x_f \geq 1$ is a valid inequality for $\mathcal{D}om\mathcal{P}_{G,K}$ (in fact this is a subset inequality, see Section 4.3), and the face $x_e = 0$ is strictly contained in the face defined by this inequality. So $x_e \geq 0$ is not a facet.

If $(G - \{e, f\})_K$ is two-connected for all $f \in E$, $f \neq e$, then, by Remark 4.2, the incidence vectors of $E \backslash \{e\}$ and $E \backslash \{e, f\}$, for all $f \neq e$, belong to $\mathcal{D}om\mathcal{P}_{G,K}$ and are affinely independent. Moreover, all of them satisfy $x_e = 0$, so the dimension of the face induced by $x_e \geq 0$ is at least $m - 1$. Since $x_e = 0$ does not hold for all $x \in \mathcal{D}om\mathcal{P}_{G,K}$, $x_e \geq 0$ defines a facet of the polytope. □

Note that the other trivial inequalities ($x_e \leq 1$) are not valid for $\mathcal{D}om\mathcal{P}_{G,K}$. However, the conditions of Proposition 4.4 are sufficient conditions under which these inequalities define facets of $\mathcal{P}_{G,K}$.

PROPOSITION 4.6 *Let $G = (V, E)$ be a graph, $K > 0$ a given constant, $G_K = (V, E_K)$ the restriction of G to bounded rings and $e \in E$ an edge of the graph. If*

1. $G = G_K$,

2. *$G - e$ is a feasible solution of 2CNBR for all $e \in E$,*

then $x_e \leq 1$ defines a facet of $\mathcal{P}_{G,K}$.

PROOF.

The incidence vectors of E and $E \backslash \{f\}$ for all $f \neq e$ belong to $\mathcal{P}_{G,K}$ and are affinely independent. Moreover, all of them satisfy $x_e = 1$, so the dimension of the face induced by $x_e \leq 1$ is at least $m - 1$. Since $x_e = 1$ does not hold for all $x \in \mathcal{P}_{G,K}$, $x_e \leq 1$ defines a facet of the polytope. □

4.2 CUT CONSTRAINTS

In this section, we examine inequalities that impose two-edge-connectivity in our formulations of 2CNBR, namely the cut constraints (3.1). We want to characterize which cut constraints define facets of $\mathcal{D}om\mathcal{P}_{G,K}$. Stoer [75] gives necessary and sufficient conditions for these constraints to define facets of the polytope of the solutions of low-connectivity constrained network design problems. If we only deal with two-connected networks, these conditions become much more simpler. The next theorem characterizes which cut constraints define facets of the polytope of two-connected networks, $\mathcal{P}_{G,\infty}$.

THEOREM 4.7 *Let $G = (V, E)$ be a graph and $W \subseteq V$ a subset of nodes, $\phi \neq W \neq V$. Then $x(\delta(W)) \geq 2$ defines a facet of $\mathcal{P}_{G,\infty}$ (and $\mathcal{D}om\mathcal{P}_{G,\infty}$) if and only if $G(W)$ and $G(V \backslash W)$ are two-connected.*

PROOF.

See Stoer [75], Theorem 5.4, page 51. □

It is easy to see that these conditions are also necessary conditions under which cut constraints define facets of $\mathcal{D}om\mathcal{P}_{G,K}$.

COROLLARY 4.8 *Let $G = (V, E)$ be a graph, $K > 0$ a given constant, and $W \subseteq V$ a subset of nodes, $\phi \neq W \neq V$. If $x(\delta(W)) \geq 2$ defines a facet of $\mathcal{D}om\mathcal{P}_{G,K}$, then the conditions of Theorem 4.7 are satisfied.*

PROOF.

If the conditions are not satisfied, then $x(\delta(W)) \geq 2$ does not define a facet of $\mathcal{D}om\mathcal{P}_{G,\infty}$. Since $\mathcal{D}om\mathcal{P}_{G,K} \subseteq \mathcal{D}om\mathcal{P}_{G,\infty}$, the cut constraint does not define a facet of $\mathcal{D}om\mathcal{P}_{G,K}$. □

Unfortunately, the conditions of Theorem 4.7 are not sufficient when we add bounded ring constraints, as illustrated by the following example. Consider the graph depicted in Figure 4.2 with unit edge lengths and a bound $K = 4$ on the ring lengths. The cut defined by $W = \{1, 2, 3, 4, 5, 6, 7\}$ satisfies the conditions of Theorem 4.7. To determine if $x(\delta(W)) \geq 2$ defines a facet, we consider the vectors of $\mathcal{D}om\mathcal{P}_{G,K}$ that lie on the face, i.e. such that $x(\delta(W)) = 2$. The pairs of edges in $\delta(W)$ whose incidence vectors have that property are

$$\begin{array}{rcl} \{4,8\} & \text{and} & \{5,9\}, \\ \{6,10\} & \text{and} & \{5,9\}, \\ \{6,10\} & \text{and} & \{7,11\}, \\ \{5,8\} & \text{and} & \{6,9\}, \\ \{7,10\} & \text{and} & \{6,9\}. \end{array}$$

We can easily conclude that all points in the face defined by $x(\delta(W)) = 2$ satisfy the following equations :

$$\begin{aligned} x_{\{4,8\}} + x_{\{6,10\}} + x_{\{5,8\}} + x_{\{7,10\}} &= 1, \\ x_{\{5,9\}} + x_{\{7,11\}} + x_{\{6,9\}} &= 1. \end{aligned}$$

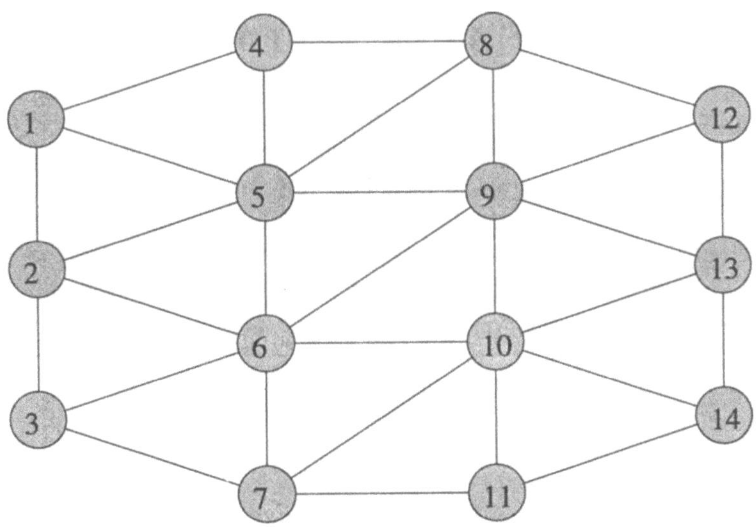

Figure 4.2. An example for which cut constraints are not facet-defining ($K = 4$)

Therefore, the dimension of the face is at most $m-2$ and $x(\delta(W)) \geq 2$ does not define a facet.

From this example, we see that we need additional conditions to characterize which cut constraints are facet-defining. For any pair of edges $e, f \in \delta(W)$, it is useful to know if there exists a vector of $\mathcal{D}om\mathcal{P}_{G,K}$ lying in the face $x(\delta(W)) = 2$ with e and f being the two edges in the cut. This is the case if and only if the incidence vector of

$$C_{e,f} := E(W) \cup E(V \backslash W) \cup \{e, f\}$$

belongs to $\mathcal{D}om\mathcal{P}_{G,K}$, i.e. if $(C_{e,f})_K$ is two-connected. A useful tool to represent and analyze the vectors belonging to the face defined by a cut constraint is the *ring-cut graph* defined below.

DEFINITION 4.2 (RING-CUT GRAPH)
Let $G = (V, E)$ be a graph, $K > 0$ a given constant, and $W \subseteq V$ a subset of nodes, $\phi \neq W \neq V$.
The ring-cut graph $RCG_{W,K} := (\delta(W), RCE_{W,K})$ induced by W is the graph defined by associating one node to each edge in $\delta(W)$ and by the set of edges

$$RCE_{W,K} = \{\{e, f\} \subseteq \delta(W) : (C_{e,f})_K \text{ is two-connected}\}.$$

The ring-cut graph corresponding to the previous example is depicted in Figure 4.3. Note that this graph is bipartite, which implies that the cut constraint does not define a facet, as we show below (see Theorem 4.10).

With the help of the ring-cut graph, we are now ready to characterize which cut constraints are facet-defining. Before giving this characterization, the following technical lemma shows that we can restrict our attention to the support of an inequality — variables having non-zero coefficients — in facet proofs for the dominant of a polytope.

LEMMA 4.9 *Let $P \subseteq \mathbb{R}^n$ be a polytope and $a^T x \geq \alpha$ and $b^T x \geq \beta$ be two valid inequalities for the dominant of P. If the face induced by $a^T x = \alpha$ is included in the face induced by $b^T x = \beta$, then*

$$a_i = 0 \Rightarrow b_i = 0, \quad i = 1, \ldots, n.$$

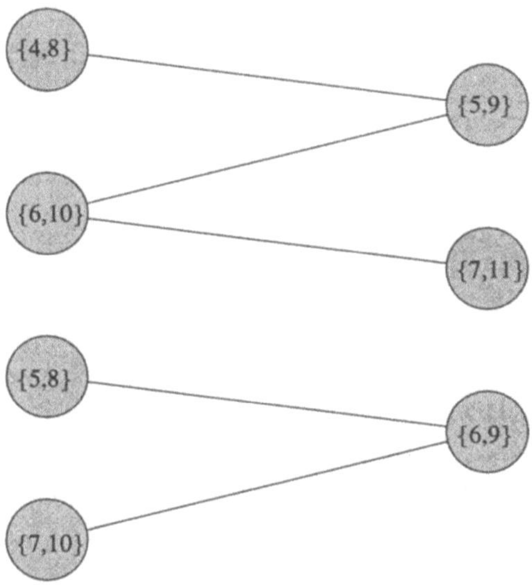

Figure 4.3. Ring-cut graph for $W = \{1,2,3,4,5,6,7\}$ ($K = 4$)

PROOF.

Let $\tilde{x}$ be a vector on the face $a^T x = \alpha$. If $a_i = 0$, then $\tilde{x} + e_i$ belongs also to the face induced by $a^T x = \alpha$. Since this face is included in the face induced by $b^T x = \beta$, $\tilde{x}$ and $\tilde{x} + e_i$ belong to the face induced by $b^T x = \beta$, implying that

$$b_i \tilde{x_i} = b_i(\tilde{x_i} + 1)$$

and thus $b_i = 0$. □

THEOREM 4.10 *Let $G = (V, E)$ be a graph, $K > 0$ a given constant, and $W \subseteq V$ a subset of nodes, $\phi \neq W \neq V$.*

$$x(\delta(W)) \geq 2$$

defines a facet of $\mathcal{D}om\mathcal{P}_{G,K}$ if and only if

1. *for all $e \in \delta(W)$, there exists $f \in \delta(W)$ such that $(C_{e,f})_K$ is two-connected;*

2. *in each connected component of $RCG_{W,K}$, there exists a cycle of odd cardinality.*

PROOF.

Necessity

If there exists $e \in \delta(W)$ such that for all $f \in \delta(W)$, $(C_{e,f})_K$ is not two-connected, then there is no feasible solution of 2CNBR, and therefore no point in $\mathcal{D}om\mathcal{P}_{G,K}$, such that $x_e = 1$ and $x(\delta(W)) = 2$. The face induced by $x(\delta(W)) \geq 2$ is thus completely included in the face defined by $x_e \geq 0$, and the cut constraint does not define a facet.

Let $S \subseteq \delta(W)$ be a connected component of $RCG_{W,K}$. If there is no cycle of odd cardinality in the subgraph $RCG_{W,K}(S)$, then this subgraph is bipartite. Therefore, there exist two subsets S_1 and S_2 such that $S_1 \cup S_2 = S$, $S_1 \cap S_2 = \phi$ and $RCE_{W,K}(S_1) = RCE_{W,K}(S_2) = \phi$. Any graph whose incidence vector belongs to $\mathcal{D}om\mathcal{P}_{G,K}$ and lies in the face $x(\delta(W)) = 2$ contains exactly two edges e and f in the cut. These edges are such that $(C_{e,f})_K$ is two-connected, there is thus one edge between e and f in $RCE_{W,K}$. Therefore, either one of these edges belongs to S_1 and the other to S_2, either the two edges belong to $\delta(W)\backslash S$. It is easy to see that this implies that the incidence vector of the graph satisfies the following system of equalities :

$$\begin{aligned} 2x(S_1) + x(\delta(W)\backslash S) &= 2, \\ 2x(S_2) + x(\delta(W)\backslash S) &= 2. \end{aligned}$$

This means that all points in the face $x(\delta(W)) = 2$ satisfy a system of equalities of dimension 2, the dimension of the face is thus at most $m - 2$ and the cut constraint does not define a facet.

Sufficiency

Let the conditions be satisfied for some inequality $a^T x := x(\delta(W)) \geq 2$. Let $b^T x \geq \beta$ be a facet-defining inequality such that the face F_a induced by $a^T x \geq 2$ in $\mathcal{D}om\mathcal{P}_{G,K}$ is contained in the face F_b induced by $b^T x \geq \beta$. Our aim is to show that b is a positive multiple of a, which implies that F_a is identical with the face F_b. From Lemma 4.9, we already know that $b_e = 0$ for all $e \in E(W) \cup E(V\backslash W)$. The remaining of the proof is divided in two steps : we first show that the coefficients of variables corresponding to edges in the same connected component of the ring-cut graph are equal, then we show that these coefficient are equal for edges belonging to different components.

1. $b_e = \gamma_S$ for all $e \in S$, for all connected component S of $RCG_{W,K}$.

We first show that b_e has the same value γ_S for all edges belonging to the same connected component S of $RCG_{W,K}$. Consider a tree rooted at k in S. In this tree, there is a unique path between k and any $l \in S$. We can partition S in two subsets S_1 and S_2 where

$$\begin{aligned} S_1 = \{l \in S : & \text{ the path from } k \text{ to } l \\ & \text{contains an odd number of edges}\}, \\ S_2 = \{l \in S : & \text{ the path from } k \text{ to } l \\ & \text{contains an even number of edges}\}. \end{aligned}$$

Let $e, f \in S_1$, $e \neq f$. By the definition of S_1, the unique path from e to f in the tree contains an even number of edges. Let $e = e_0, e_1, \ldots, e_{2j} = f$ be this path. By the definition of the ring-cut graph, the existence of this path means that $(C_{e_i,e_{i+1}})_K$ and $(C_{e_{i+1},e_{i+2}})_K$ are two-connected for $i = 0, \ldots, 2(j-1)$. Therefore, $b_{e_i} = b_{e_{i+2}}$ and

$$b_e = b_{e_0} = b_{e_2} = \ldots = b_{e_{2j}} = b_f.$$

Thus all edges in S_1 have the same coefficient γ_{S_1}. The same reasoning applies to S_2, and all edges in S_2 have the same coefficient γ_{S_2}.

Since there exists a cycle of odd cardinality in S, there must exist an edge between two nodes in S_1 or between two nodes in S_2. Without loss of generality, suppose there exists an edge between e and f in S_1, i.e. $(C_{e,f})_K$ is two-connected. Moreover, since S is connected, there exists an edge between some $g \in S_1$ and some $h \in S_2$, and $(C_{g,h})_K$ is two-connected. Thus, $b_e + b_f = b_g + b_h$. Since $b_e = b_f = b_g = \gamma_{S_1}$ and $b_h = \gamma_{S_2}$, we can conclude that

$$\gamma_{S_1} = \gamma_{S_2} = \gamma_S$$

and all edges in S have the same coefficient.

2. $\gamma_S = \gamma_T$ **for all connected components** S, T **of** $RCG_{W,K}$.

To prove that all edges in $\delta(W)$ have the same coefficient in the inequality, it remains to prove that the coefficients are equal for two edges belonging to different connected components S and T of $RCG_{W,K}$. For all $e \in S$, there exists f such that $(C_{e,f})_K$ is two-connected. Therefore, $f \in S$ and $b_e + b_f = 2\gamma_S$. Similarly, we can find $g, h \in T$ such that $(C_{g,h})_K$ is two-connected and $b_g + b_h = 2\gamma_T$. So, $b_e + b_f = b_g + b_h$ and we can conclude that $\gamma_S = \gamma_T = \gamma$.

We have proved that $b = \gamma a$. Since F_b cannot define a facet if $b \leq 0$, we have $\gamma > 0$. Thus $x(\delta(W)) \geq 2$ and $b^T x \geq \beta$ define the same facet $F_a = F_b$. □

4.3 NODE-CUT AND SUBSET INEQUALITIES

After examining inequalities that impose two-edge-connectivity, we study in this section node-cut constraints (3.2), i.e. constraints that impose two-node-connectivity. Again, Stoer [75] gives necessary and sufficient conditions for these constraints to define facets of the polytope of the solutions of low-connectivity constrained network design problems. If we only deal with two-connected networks, these conditions become much more simpler. The next theorem characterizes which node-cut constraints define facets of the polytope of two-connected networks, $\mathcal{P}_{G,\infty}$.

THEOREM 4.11 *Let* $G = (V, E)$ *be a graph,* $z \in V$ *be a node of the graph and* $W \subseteq V \backslash \{z\}$ *a subset of nodes,* $\phi \neq W \neq V$. *Denote by* $\overline{W}$ *the set* $V \backslash (W \cup \{z\})$. *Then* $x(\delta_{G-z}(W)) \geq 1$ *defines a facet of* $\mathcal{P}_{G,\infty}$ *(and* $\mathcal{D}om\mathcal{P}_{G,\infty}$*) if and only if*

1. $G(W)$ *and* $G(\overline{W})$ *are two-edge-connected;*
2. $G(W \cup \{z\})$ *and* $G(\overline{W} \cup \{z\})$ *are two-node-connected.*

PROOF.

See Stoer [75], Theorem 5.16, page 63. □

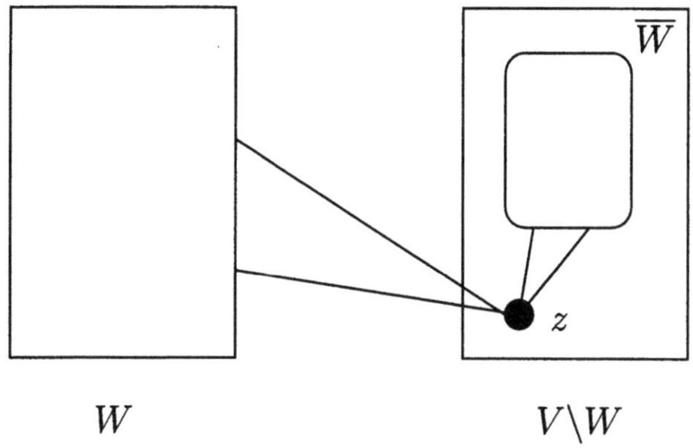

Figure 4.4. $G - (\delta(W)\backslash\delta(\{z\}))$

As for cut constraints, it is easy to see that these conditions are also necessary conditions under which node-cut constraints define facets of $\mathcal{D}om\mathcal{P}_{G,K}$, but that they are not sufficient. To derive sufficient conditions, we give here another interpretation of node-cut constraints, and show that they are a subset of a larger class of facet-defining inequalities.

Consider the cut $\delta(W)$ defined by a subset $W \subseteq V$ of nodes and a node $z \in V\backslash W$.

If S is the subset of edges not incident to z in the cut, i.e. $S = \delta_{G-z}(W) = \delta(W)\backslash\delta(\{z\})$, it is clear that $G - S$ is not two-connected, since z is an articulation point in this graph, as illustrated in Figure 4.4. Therefore, there must be one edge in $S = \delta_{G-z}(W)$, leading to the node-cut constraint $x(\delta_{G-z}(W)) \geq 1$.

A more general result can be established for any subset $S \subseteq E$: if $(G - S)_K$ is not two-connected, then S must contain one edge in order to obtain a feasible solution. This leads us to new valid inequalities.

PROPOSITION 4.12 *Let $G = (V, E)$ be a graph, $K > 0$ a given constant, and $S \subseteq E$ a subset of edges.*

If $(G-S)_K$ is not two-connected, then

$$x(S) \geq 1 \tag{4.1}$$

is a valid inequality for the 2CNBR problem.

PROOF.

Let F be a feasible solution to the 2CNBR problem and x be its incidence vector. Since $(G-S)_K$ is not two-connected, $G-S$ does not contain a feasible solution. Therefore, F contains at least one edge belonging to S. We can conclude that (4.1) are valid inequalities for the 2CNBR problem. □

Inequalities (4.1) are called *subset inequalities*. The next theorem shows under which conditions these inequalities define facets of $\mathcal{D}om\mathcal{P}_{G,K}$.

THEOREM 4.13 *Let $G=(V,E)$ be a graph, $K>0$ a given constant, and $S \subseteq E$ a subset of edges such that $(G-S)_K$ is not two-connected. Then*

$$x(S) \geq 1$$

defines a facet of $\mathcal{D}om\mathcal{P}_{G,K}$ if and only if for all $e \in S$, $(G-(S\backslash\{e\}))_K$ is two-connected.

PROOF.

Necessity

Suppose that for some $e \in S$, $(G-(S\backslash\{e\}))_K$ is not two-connected. Then there is no feasible solution such that $x(S)=1$ and $x_e=1$, thus the face induced by the subset inequality is completely included in the face defined by $x_e \geq 0$, and the subset inequality does not define a facet.

Sufficiency

Let the conditions be satisfied for some subset inequality

$$a^T x := x(S) \geq 1.$$

Let $b^T x \geq \beta$ be a facet-defining inequality such that the face F_a induced by $a^T x \geq 1$ in $\mathcal{D}om\mathcal{P}_{G,K}$ is contained in the facet F_b induced by $b^T x \geq \beta$. Our aim is to show that b is a positive multiple of a, which implies that F_a is identical with the facet F_b. From Lemma 4.9, we already know that $b_e = 0$ for all $e \in E \backslash S$.

It remains to show that b_e has the same value γ for all $e \in S$. Since for all $e \in S$, $(G - (S \backslash \{e\}))_K$ is two-connected, for any $e, f \in S$, $e \neq f$, the incidence vectors of $(G - (S \backslash \{e\}))$ and $(G - (S \backslash \{f\}))$ belong to $\mathcal{D}om\mathcal{P}_{G,K}$. Therefore, $b_e = b_f = \gamma$.

So we have proved that $b = \gamma a$. Since F_b cannot define a facet if $b \leq 0$, we have $\gamma > 0$. Thus $x(S) \geq 1$ and $b^T x \geq \beta$ define the same facet $F_a = F_b$. □

Consider again the graph depicted in Figure 4.2, with $K = 4$. The node-cut constraint defined by $W = \{1, 2, 3, 4, 5, 6, 7\}$ and $z = 8$,

$$x_{\{5,9\}} + x_{\{6,9\}} + x_{\{6,10\}} + x_{\{7,10\}} + x_{\{7,11\}} \geq 1,$$

does not define a facet since it is induced by

$$S = \{\{5, 9\}, \{6, 9\}, \{6, 10\}, \{7, 10\}, \{7, 11\}\}$$

and $(G - (S \backslash \{\{7, 11\}\}))_4$ is not two-connected. Therefore,

$$S = \{\{5, 9\}, \{6, 9\}, \{6, 10\}, \{7, 10\}\}$$

induces a stronger subset inequality,

$$x_{\{5,9\}} + x_{\{6,9\}} + x_{\{6,10\}} + x_{\{7,10\}} \geq 1.$$

This inequality is facet-defining since adding any edge $e \notin S$ to S induces a two-connected $(G - (S \backslash \{e\}))_K$.

This shows that subset inequalities are a useful generalization of node-cut inequalities for the 2CNBR problem. Note that subset inequalities are also valid for two-connected networks (without ring constraints). However, in this case, the only interesting subset inequalities (i.e. those that are facet-defining) are node-cut constraints, as shown in the next proposition.

PROPOSITION 4.14 *Let $G = (V, E)$ be a graph, and $S \subseteq E$ a subset of edges such that $G - S$ is not two-connected.*

If $x(S) \geq 1$ defines a facet of $\mathcal{P}_{G,\infty}$ (and $Dom\mathcal{P}_{G,\infty}$), then this inequality is a node-cut constraint.

PROOF.

Since $G - S$ is not two-connected, it contains at least one articulation point $z \in V$. Let $W \subseteq V$ be a subset of nodes defining a connected component of $(G - S) - z$. Since $x(S) \geq 1$ defines a facet, by Theorem 4.13, for any $e \in S$, $G - (S \backslash \{e\})$ is two-connected. Therefore, at least one edge of $\delta(W) \backslash \delta(\{z\})$ belongs to $G - (S \backslash \{e\})$. The only possible edge is e, thus $e \in \delta(W) \backslash \delta(\{z\})$.

We can conclude that $S \subseteq \delta(W) \backslash \delta(\{z\})$.

If the two sets are equal, then

$$S = \delta_{G-z}(W)$$

and the subset inequality is a node-cut constraint.

Otherwise, there exists some edge $e \in \delta(W) \backslash S$ which is not incident to z, as illustrated in Figure 4.5. Since z is an articulation point in $G - S$, there exists a subset $\overline{W} \subseteq V$ such that $\delta_{(G-S)-z}(\overline{W}) = \phi$. Let $W_1 = W \backslash \overline{W}$. It is easy to see that $\delta(W \backslash W_1)$ must contain all the edges of $\delta(W) \backslash \delta(\{z\})$ that are not in S, and that there is no edge between W_1 and $W \backslash W_1$ in G. Therefore,

$$\delta_{G-z}(W_1) \subseteq \delta(W) \backslash (\delta(\{z\}) \cup \delta(W \backslash W_1)) \subseteq S.$$

The node-cut constraint $x(\delta_{G-z}(W_1)) \geq 1$ is thus stronger than the subset inequality $x(S) \geq 1$. Since this last inequality is facet-defining, the two inequalities are the same, and the subset inequality is a node-cut constraint. □

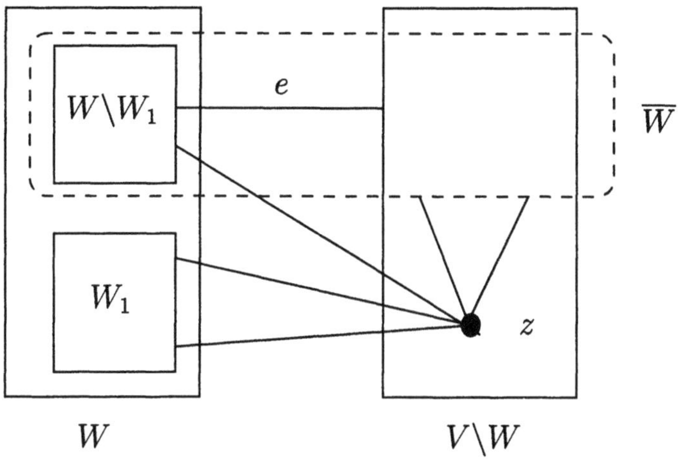

Figure 4.5. A graph such that $G - S$ is not two-connected and $S \neq \delta(W)\backslash\delta(\{z\})$

4.4 RING-CUT INEQUALITIES

The ring-cut graph introduced in Section 4.2 allows us to determine if a cut constraint is facet-defining. In this section, we use the ring-cut graph to generate new valid inequalities for the 2CNBR problem.

Consider the graph depicted in Figure 4.6 with unit edge lengths and a bound $K = 4$ on the ring lengths. The ring-cut graph corresponding to the cut defined by $W = \{1, 2, 3, 4, 5, 6\}$ is depicted in Figure 4.7. Since this graph is bipartite, it follows from Theorem 4.10 that the cut constraint $x(\delta(W)) \geq 2$ does not define a facet. Our aim is to find some valid inequality that dominates the cut constraint. The cut constraint is not facet-defining because there are not enough edges in the ring-cut graph, i.e. not enough pairs of edges $e, f \in \delta(W)$ such that $(C_{e,f})_K$ is two-connected. A way to find a stronger inequality is to extend the search for feasible points to solutions using more than two edges in the cut.

To do so, we extend the notation $C_{e,f}$ to any subset $S \subseteq \delta(W)$ by defining

$$C_S := E(W) \cup E(V\backslash W) \cup S.$$

For the example of Figure 4.6, we can easily enumerate all the minimal subsets $S \subseteq \delta(W)$ such that $(C_S)_K$ is two-connected. These subsets are :

$$\{\{3,7\},\{4,8\}\},$$
$$\{\{5,7\},\{4,8\}\},$$
$$\{\{5,9\},\{4,8\}\},$$
$$\{\{5,9\},\{6,10\}\},$$
$$\{\{3,7\},\{5,7\},\{5,9\}\}.$$

It is easy to see that the incidence vectors of these subsets satisfy

$$x_{\{3,7\}} + x_{\{5,7\}} + x_{\{5,9\}} + 2x_{\{4,8\}} + 2x_{\{6,10\}} = 3.$$

Moreover, a subset $\overline{S} \subseteq \delta(W)$ induces a two-connected $(C_{\overline{S}})_K$ if and only if $\overline{S}$ contains one of the above subsets. Therefore, any feasible solution satisfies the inequality

$$x_{\{3,7\}} + x_{\{5,7\}} + x_{\{5,9\}} + 2x_{\{4,8\}} + 2x_{\{6,10\}} \geq 3.$$

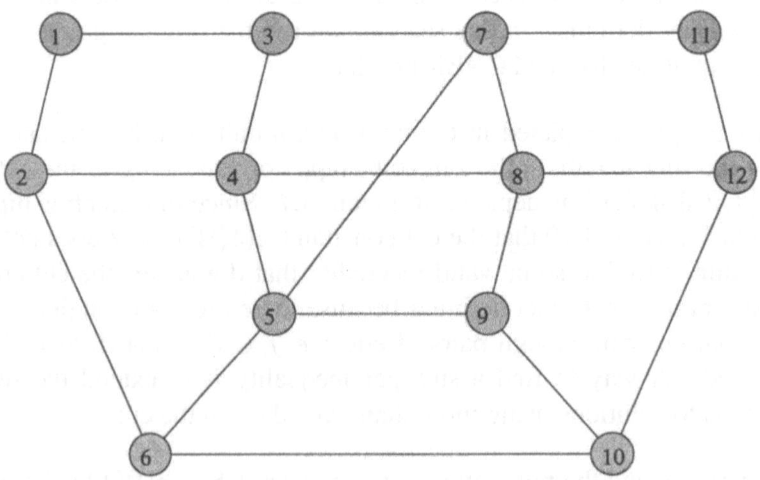

Figure 4.6. Graph for which some ring-cut inequalities are valid

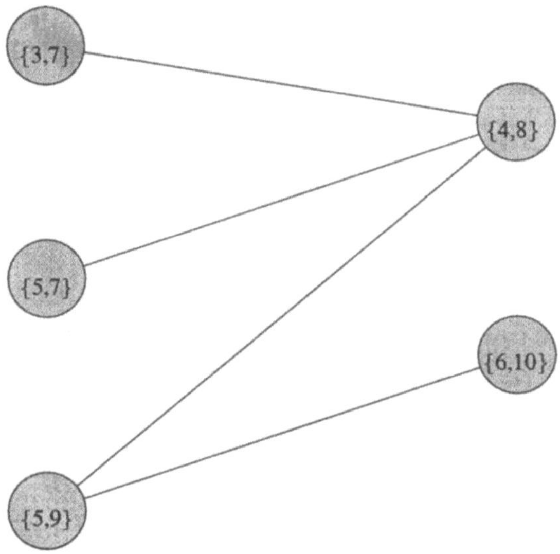

Figure 4.7. The ring-cut graph for $W = \{1, 2, 3, 4, 5, 6\}$

The next proposition generalizes this result.

PROPOSITION 4.15 *Let $G = (V, E)$ be a graph, $K > 0$ a given constant, $W \subseteq V$ a subset of nodes, $\phi \neq W \neq V$.*

If $S \subseteq \delta(W)$ is an independent subset in the ring-cut graph $RCG_{W,K}$, then

$$x(S) + 2x(\delta(W)\backslash S) \geq 3 \qquad (4.2)$$

is a valid inequality for the 2CNBR problem.

PROOF.

Let F be a feasible solution to the 2CNBR problem and x be its incidence vector. Since F is two-connected, $x(\delta(W)) \geq 2$ and therefore,

$$x(S) + 2x(\delta(W)\backslash S) \geq 2.$$

If this inequality is satisfied as equality by F, then F contains exactly two edges e and f in $\delta(W)$. These edges belong to S since edges in S are the only edges having a coefficient equal to 1 in the inequality. Moreover, $(C_{e,f})_K$ is two-connected since it includes F, thus there is an edge between e and f in the ring-cut graph. Since S is an independent subset, this leads to a contradiction. We can conclude that (4.2) is a valid inequality for the 2CNBR problem. □

Inequalities (4.2) are called ring-cut inequalities. Note that if the ring-cut graph is bipartite, as is the case in Figure 4.7, then it is possible to partition $\delta(W)$ in two independent subgraphs S_1 and S_2 that define the two ring-cut inequalities

$$\begin{aligned} x(S_1)+ \quad 2x(S_2) &\geq 3, \\ 2x(S_1)+ \quad x(S_2) &\geq 3. \end{aligned}$$

The sum of these inequalities, divided by 3, gives the cut constraint $x(\delta(W)) \geq 2$. Therefore, in this case, ring-cut inequalities are stronger than cut constraints.

We were not able to characterize completely which ring-cut inequalities are facet-defining. However, we present here a set of necessary conditions, as well as two sets of sufficient conditions.

THEOREM 4.16 *Let $G = (V, E)$ be a graph, $K > 0$ a given constant, $W \subseteq V$ a subset of nodes, $\phi \neq W \neq V$.*

If $S \subseteq \delta(W)$ is an independent subset in the ring-cut graph $RCG_{W,K}$ and if

$$x(S) + 2x(\delta(W)\backslash S) \geq 3$$

defines a facet of $\mathcal{D}om\mathcal{P}_{G,K}$, then

1. *for all $e \in S$, either there exists $f \in \delta(W)\backslash S$ such that $(C_{e,f})_K$ is two-connected or there exist $g, h \in S$ such that $(C_{e,g,h})_K$ is two-connected;*
2. *for all $e \in \delta(W)\backslash S$, there exists $f \in S$ such that $(C_{e,f})_K$ is two-connected;*
3. *there exist $e, f, g \in S$ such that $(C_{e,f,g})_K$ is two-connected.*

PROOF.

First, note that the incidence vector of a graph lies in the face defined by (4.2) if and only if the graph contains three edges belonging to S and no edge in $\delta(W)\backslash S$, or one edge in S and one edge in $\delta(W)\backslash S$.

Suppose one of the first two conditions is not satisfied for some $e \in \delta(W)$. It is easy to see from the remark above that in this case, there is no vector of $\mathcal{D}om\mathcal{P}_{G,K}$ with $x_e = 1$ lying in the face defined by (4.2). Therefore, this face is completely included in the face defined by $x_e \geq 0$ and (4.2) does not define a facet.

If the third condition is not satisfied, then any graph whose incidence vector lies in the face defined by (4.2) contains exactly one edge in S and one edge in $\delta(W)\backslash S$. Therefore, these vectors satisfy the following system of equalities :

$$\begin{aligned} x(S) &= 1, \\ x(\delta(W)\backslash S) &= 1. \end{aligned}$$

This means that all points in the face defined by (4.2) satisfy a system of equalities of dimension 2, thus the dimension of the face is at most $m - 2$ and (4.2) does not define a facet. □

Consider again the example of Section 4.2. The ring-cut graph depicted in Figure 4.3 is bipartite, and

$$\begin{aligned} &x_{\{4,8\}} + x_{\{6,10\}} + x_{\{5,8\}} + x_{\{7,10\}} \\ &\quad +2x_{\{5,9\}} + 2x_{\{7,11\}} + 2x_{\{6,9\}} \geq 3 \end{aligned}$$

is a valid ring-cut inequality. However, it is easy to check that the third condition of Theorem 4.16 is not satisfied, and therefore the inequality is not facet-defining.

THEOREM 4.17 *Let $G = (V, E)$ be a graph, $K > 0$ a given constant, $W \subseteq V$ a subset of nodes, $\phi \neq W \neq V$.*

If $S \subseteq \delta(W)$ is an independent subset in the ring-cut graph $RCG_{W,K}$, and if

1. the bipartite subgraph $(\delta(W), \delta_{RCG_{W,K}}(S))$ of $RCG_{W,K}$ — obtained by considering only edges leaving S — is connected;

2. there exist $e, f, g \in S$ such that $(C_{e,f,g})_K$ is two-connected;

then

$$x(S) + 2x(\delta(W)\backslash S) \geq 3$$

defines a facet of $\mathcal{D}om\mathcal{P}_{G,K}$.

PROOF.

Let the conditions be satisfied for some inequality $a^T x := x(S) + 2x(\delta(W)\backslash S) \geq 3$. Let $b^T x \geq \beta$ be a facet-defining inequality such that the face F_a induced by $a^T x \geq 3$ in $\mathcal{D}om\mathcal{P}_{G,K}$ is contained in the face F_b induced by $b^T x \geq \beta$. Our aim is to show that b is a positive multiple of a, which implies that F_a is identical with the face F_b. From Lemma 4.9, we already know that $b_e = 0$ for all $e \in E(W) \cup E(V\backslash W)$. The remaining of the proof is divided in three steps : we first show that the coefficients of variables corresponding to edges belonging to S are equal, a similar proof leads to the same result for edges in $\delta(W)\backslash S$, then we establish the relationship between all the coefficients.

1. $b_e = \gamma_S$ for all $e \in S$.

We first show that b_e has the same value γ_S for all edges belonging to S. Consider two edges $e, f \in S$, $e \neq f$. Since $(\delta(W), \delta_{RCG_{W,K}}(S))$ is connected and bipartite, there exists a path of even cardinality from e to f in $RCG_{W,K}$. By arguments similar to those of Theorem 4.10, we can conclude that $b_e = b_f$.

2. $b_e = \overline{\gamma_S}$ for all $e \in \delta(W)\backslash S$.

A similar proof leads to the fact that b_e has the same value $\overline{\gamma_S}$ for all edges belonging to $\delta(W)\backslash S$.

3. $\overline{\gamma_S} = 2\gamma_S$.

There exist $e, f, g \in S$ such that $(C_{e,f,g})_K$ is two-connected. Moreover, since $(\delta(W), \delta_{RCG_{W,K}}(S))$ is connected, there exists $h \in \delta(W)\backslash S$ such that $(C_{e,h})_K$ is two-connected. Therefore, $b_e + b_f + b_g = b_e + b_h$. Since $b_e = b_f = b_g = \gamma_S$ and $b_h = \overline{\gamma_S}$, we can conclude that $\overline{\gamma_S} = 2\gamma_S$.

We have proved that $b = \gamma_S a$. Since F_b cannot define a facet if $b \leq 0$, we have $\gamma_S > 0$. Thus $x(S) + 2x(\delta(W)\backslash S) \geq 3$ and $b^T x \geq \beta$ define the same facet $F_a = F_b$. □

The example of ring-cut inequality given at the beginning of this section satisfies these conditions, and therefore is facet-defining.

Another interesting case arises when the ring-cut graph contains no edge. Then $S = \delta(W)$ is independent and the ring-cut inequality becomes

$$x(\delta(W)) \geq 3.$$

The next theorem gives sufficient conditions for these inequalities to be facet-defining.

THEOREM 4.18 *Let $G = (V, E)$ be a graph, $K > 0$ a given constant, $W \subseteq V$ a subset of nodes, $\phi \neq W \neq V$.*

If $RCE_{W,K} = \phi$ and if, for all $e, f \in \delta(W)$, there exist $g, h \in \delta(W)\backslash\{e, f\}$ such that $(C_{e,g,h})_K$ and $(C_{f,g,h})_K$ are two-connected, then $x(\delta(W)) \geq 3$ defines a facet of $\mathcal{D}om\mathcal{P}_{G,K}$.

PROOF.

Let the conditions be satisfied for some inequality $a^T x := x(\delta(W)) \geq 3$. Let $b^T x \geq \beta$ be a facet-defining inequality such that the face F_a induced by $a^T x \geq 3$ in $\mathcal{D}om\mathcal{P}_{G,K}$ is contained in the face F_b induced by $b^T x \geq \beta$. Our aim is to show that b is a positive multiple of a, which implies that F_a is identical with the face F_b. From Lemma 4.9, we already know that $b_e = 0$ for all $e \in E(W) \cup E(V\backslash W)$.

For any $e, f \in \delta(W)$, there exist $g, h \in \delta(W)\backslash\{e, f\}$ such that $(C_{e,g,h})_K$ and $(C_{f,g,h})_K$ are two-connected. Therefore $b_e = b_f$ and all edges in $\delta(W)$ have the same coefficient γ in the inequality.

We have proved that $b = \gamma a$. Since F_b cannot define a facet if $b \leq 0$, we have $\gamma > 0$. Thus $x(\delta(W)) \geq 3$ and $b^T x \geq \beta$ define the same facet $F_a = F_b$. □

This theorem has useful implications for particular cases of the 2CNBR problem, as we will see in Section 5.4.

4.5 RING-COVER INEQUALITIES

We now study inequalities imposing that each edge belongs to a feasible ring. In this section, we introduce inequalities coming from the formulation (3.8)-(3.13), where we enumerate all the feasible cycles.

Suppose we have a two-connected solution, with an (infeasible) edge e violating the ring constraints (3.10) and (3.11), i.e. at least one edge in each feasible cycle $c \in \mathcal{C}_e$ is not in the current solution. A valid inequality can be generated by imposing that at least one of these edges must be in the solution, as proposed by Maffioli and Värbrand [65].

We provide here another way of deriving these inequalities by looking at the projection of our formulation of ring constraints. Suppose $\tilde{\mathbf{x}} = (\tilde{x}_e)_{e \in E}$ is a vector not satisfying ring constraints (3.10)-(3.11), i.e. such that the linear system

$$\begin{aligned} &\sum_{c \in \mathcal{C}_e:\, f \in c} y_e^c \leq \tilde{x}_f && e \in E,\ f \in E \backslash \{e\}, \\ &\sum_{c \in \mathcal{C}_e} y_e^c \geq \tilde{x}_e && e \in E, \\ &y_e^c \geq 0 && e \in E,\ c \in \mathcal{C}_e \end{aligned}$$

is infeasible. Since $\tilde{\mathbf{x}}$ is fixed, the system can be decomposed by edge. Therefore, there exists $e \in E$ such that

$$\sum_{c \in \mathcal{C}_e} y_e^c < \tilde{x}_e$$

for all $(y_e^c)_{c \in \mathcal{C}_e}$ satisfying

$$\sum_{c \in \mathcal{C}_e:\, f \in c} y_e^c \leq \tilde{x}_f, \qquad f \in E \backslash \{e\}, \tag{4.4}$$

$$y_e^c \geq 0 \qquad\qquad c \in \mathcal{C}_e. \tag{4.5}$$

By duality, the linear program

$$\begin{array}{lll} \min & \sum\limits_{f\in E\backslash\{e\}} u_f\tilde{x_f} & \\ \text{s.t.} & \sum\limits_{f\in c\backslash\{e\}} u_f \geq 1 & c \in \mathcal{C}_e, \qquad (4.6) \\ & u_f \geq 0 & f \in E\backslash\{e\} \qquad (4.7) \end{array}$$

has an optimal value smaller that $\tilde{x_e}$. Let $(\overline{u}_f)_{f\in E\backslash\{e\}}$ be the optimal solution of this linear program.

A valid inequality for the 2CNBR problem is

$$\sum_{f\in E\backslash\{e\}} \overline{u}_f x_f \geq x_e. \tag{4.7}$$

We call inequalities (4.7) *ring-cover inequalities.*

Unfortunately, solving the separation problem exactly requires the knowledge of all the feasible cycles $c \in \mathcal{C}_e$.

4.6 METRIC INEQUALITIES

In this section, we describe a class of valid inequalities that arises from the projection of the flow formulation (3.14)-(3.21) of the 2CNBR problem. This approach is related to the work of Bienstock et al. [10].

PROPOSITION 4.19 *Let $G = (V, E)$ be a graph and $K > 0$ a given constant. Consider an edge $e := \{i, j\} \in E$ and a set of node potentials $(\alpha_k)_{k\in V}$ satisfying*

$$\alpha_i - \alpha_j > K - d(i, j).$$

Then

$$\sum_{f\in E\backslash\{e\}} v_f x_f \geq x_e \tag{4.8}$$

is a valid inequality for 2CNBR where

$$v_f = \min\left(1, \max\left(0, \frac{|\alpha_l - \alpha_k| - d(k,l)}{\alpha_i - \alpha_j + d(i,j) - K}\right)\right) \quad (4.9)$$

for all $f := \{k,l\} \in E \backslash \{e\}$.

PROOF.

Let $F \subseteq E$ be a feasible solution defined by the $(x_e)_{e \in E}$ variables, and consider an edge $e := \{i,j\} \in F$. This edge belongs to a feasible cycle of length less than or equal to K. It means that inequalities (3.16)-(3.20) from the flow formulation are satisfied for edge e. In other words, the problem

$$\min w(e) = \sum_{\{k,l\} \in E} d(k,l)(u_{kl} + u_{lk})$$

s.t.

$$\sum_{l:\{k,l\} \in E} (u_{kl} - u_{lk}) = \begin{cases} 2x_{\{i,j\}} & \text{if } k = i \\ -2x_{\{i,j\}} & \text{if } k = j \\ 0 & \text{otherwise} \end{cases}$$

$$k \in V, \quad (4.11)$$

$$u_{kl} + u_{lk} \leq x_{\{k,l\}} \quad \{k,l\} \in E, \quad (4.12)$$

$$u_{kl} \geq 0 \quad \{k,l\} \in E, \quad (4.13)$$

$$u_{lk} \geq 0 \quad \{k,l\} \in E \quad (4.14)$$

has an optimal solution value $w^*(e)$ such that $w^*(e) \leq K$.

On the other hand, if $\{i,j\} \notin F$, $u_{kl} = 0$, $k,l \in V$ is feasible, meaning the optimal solution of the problem is $w^*(e) = 0$. We can conclude that the optimal solution value always satisfies

$$w^*(e) \leq K\, x_{\{i,j\}}.$$

The dual of this minimum cost flow problem is

$$\max w(e) = 2(\alpha_i - \alpha_j)\, x_{\{i,j\}} - \sum_{\{k,l\} \in E} \beta_{\{k,l\}} x_{\{k,l\}}$$

s.t.

$$\alpha_k - \alpha_l - \beta_{\{k,l\}} \le d(k,l) \qquad \{k,l\} \in E, \tag{4.15}$$
$$\alpha_l - \alpha_k - \beta_{\{k,l\}} \le d(k,l) \qquad \{k,l\} \in E, \tag{4.16}$$
$$\beta_{\{k,l\}} \ge 0 \qquad \{k,l\} \in E. \tag{4.17}$$

It is easy to see that the optimum of the dual is reached for

$$\beta_{\{k,l\}} = \max\,(0, |\alpha_l - \alpha_k| - d(k,l)), \quad \{k,l\} \in E,$$

and the dual problem becomes

$$\max w(e) = (\alpha_i - \alpha_j + d(i,j))\; x_{\{i,j\}} \\ - \sum_{\{k,l\} \in E \setminus \{e\}} \max\,(0, |\alpha_l - \alpha_k| - d(k,l))\; x_{\{k,l\}}.$$

Since — by duality — for any values of $(\alpha_k)_{k \in V}$, we must have $w(e) \le w^*(e) \le K\; x_{\{i,j\}}$,

$$\sum_{\{k,l\} \in E \setminus \{e\}} \max\,(0, |\alpha_l - \alpha_k| - d(k,l))\; x_{\{k,l\}} \\ \ge (\alpha_i - \alpha_j + d(i,j) - K)\; x_{\{i,j\}}$$

is a valid inequality. Dividing by the positive coefficient of $x_{\{i,j\}}$, this leads to

$$\sum_{\{k,l\} \in E \setminus \{e\}} \max\left(0, \frac{|\alpha_l - \alpha_k| - d(k,l)}{\alpha_i - \alpha_j + d(i,j) - K}\right)\; x_{\{k,l\}} \ge x_{\{i,j\}}.$$

Since the variables are binary, a stronger valid inequality is obtained by changing the coefficient of a variable to 1 when

$$\frac{|\alpha_l - \alpha_k| - d(k,l)}{\alpha_i - \alpha_j + d(i,j) - K} > 1,$$

leading to (4.8). □

Inequalities (4.8) are called *metric inequalities*. The next proposition shows that the linear relaxation obtained with metric inequalities is weaker than the one obtained with ring-cover inequalities.

PROPOSITION 4.20 *Any metric inequality is a ring-cover inequality.*

PROOF.

To prove this result, it is sufficient to show that $(v_f)_{f \in E}$ is a feasible solution to the separation problem (4.5)-(4.6) for ring-cover inequalities. The non-negativity constraints are obviously satisfied.

Let $c \in C_e$ be a feasible cycle. If there exists $f \in c$ such that $v_f = 1$, obviously $\sum_{f \in c \setminus \{e\}} v_f \geq 1$ and (4.5) is satisfied. Otherwise, for all $f \in c$,

$$v_f = \max\left(0, \frac{|\alpha_l - \alpha_k| - d(k,l)}{\alpha_i - \alpha_j + d(i,j) - K}\right),$$

and

$$\begin{aligned}
&\sum_{f \in c \setminus \{e\}} v_f \\
&\geq \frac{\sum_{f:=\{k,l\} \in c \setminus \{e\}} (|\alpha_l - \alpha_k| - d(k,l))}{\alpha_i - \alpha_j + d(i,j) - K} \\
&\geq \frac{\sum_{f:=\{k,l\} \in c \setminus \{e\}} (\alpha_l - \alpha_k - d(k,l))}{\alpha_i - \alpha_j + d(i,j) - K} \\
&= \frac{\alpha_i - \alpha_j - \sum_{f \in c \setminus \{e\}} d_f}{\alpha_i - \alpha_j + d(i,j) - K} \\
&\geq 1,
\end{aligned}$$

since c is feasible i.e. $\sum_{f \in c \setminus \{e\}} d_f \leq K - d(i,j)$.

Constraint (4.5) is therefore satisfied. □

Despite this result, metric inequalities are easier to separate, as we will see in Section 6.3, and we only implemented the separation for metric inequalities in our Branch-and-Cut algorithm.

4.7 NODE-PARTITION INEQUALITIES

The polytope of connected networks is completely described by *partition inequalities* (Grötschel and Monma [47]). Given a partition $W_1, W_2, \ldots, W_p$ $(p \geq 2)$ of V into p nonempty subsets, the corresponding partition inequality is

$$\frac{1}{2}\sum_{i=1}^{p} x(\delta(W_i)) \geq p-1.$$

Since a two-connected network remains connected when a node is removed,

$$\frac{1}{2}\sum_{i=1}^{p} x(\delta_{G-z}(W_i)) \geq p-1$$

is a valid inequality for the polytope of two-connected networks, and therefore for 2CNBR, where $W_1, W_2, \ldots, W_p$ $(p \geq 2)$ is a partition of $V \backslash \{z\}$. These inequalities are called *node-partition inequalities.*

The following theorem gives sufficient conditions for node-partition inequalities to define facets of $\mathcal{D}om\mathcal{P}_{G,K}$.

THEOREM 4.21 *Let $G = (V, E)$ be a graph, $K > 0$ a given constant, $z \in V$ a node of the graph and $W_1, W_2, \ldots, W_p$ $(p \geq 2)$ a partition of $V \backslash \{z\}$. Let $\hat{G}$ denote the graph $(G_K - z)/W_1/\ldots/W_p$. If*

1. *$\hat{G}$ is two-connected;*
2. *$(G(W_i \cup \{z\}) - e)_K$ is two-connected for all $e \in G(W_i \cup \{z\})$;*
3. *$\{z, i, j\}$ form a feasible cycle for all $e := \{i, j\} \in \delta_{G-z}(W_i)$, $i = 1, \ldots, p$;*
4. *there exists a feasible cycle using z and f in $G - e$ for all $f \in \delta_{G-z}(W_i)$, $i = 1, \ldots, p$, and for all $e \in \delta(\{z\})$,*

then

$$a^T x := \frac{1}{2}\sum_{i=1}^{p} x(\delta_{G-z}(W_i)) \geq p-1$$

defines a facet of $\mathcal{D}om\mathcal{P}_{G,K}$.

PROOF.

Let the conditions be satisfied. We show that there are $|E|$ affinely independent vectors in the face defined by the node-partition inequality.

Let E' be the set of all edges whose coefficients in $a^T x \geq p - 1$ are 0. By the first condition, the graph $\hat{G} = (\hat{V}, \hat{E})$ contains $|\hat{E}|$ spanning trees whose incidence vectors are affinely independent (see Theorem 4.10 in [47]). Any such tree of $\hat{G}$ can be augmented by E' to a set $C \subseteq E$ whose incidence vector belongs to $\mathcal{D}om\mathcal{P}_{G,K}$. To prove this, we will show that $C = C_K$. By Theorem 6.3 in [49], the two first conditions imply that C is two-connected, and therefore the incidence vector of $C = C_K$ belongs to $\mathcal{D}om\mathcal{P}_{G,K}$.

Consider an edge $e := \{i, j\} \in C$ arising from the tree in $\hat{G}$. By the third condition, $\{z, i, j\}$ form a feasible cycle and $e \in C_K$. If $e \in E'$, then $e \in C_K$ by the second condition. Therefore, $C \subseteq C_K$. The reverse inclusion is obvious.

The same proof applies to show that the incidence vector of $C \backslash \{e\}$ belongs to $\mathcal{D}om\mathcal{P}_{G,K}$ for any $e \in E' \backslash \delta(\{z\})$. It is thus possible to construct $|E| - |\delta(\{z\})|$ independent vector lying in the face $a^T x = p - 1$. Moreover, all these vectors satisfy $x_e = 1$ for all $e \in \delta(\{z\})$.

Consider $e := \{w_i, z\} \in \delta(\{z\})$, with $w_i \in W_i$. We construct a tree $\hat{T}$ in $\hat{G}$, corresponding to a subset of edges T in E, such that $(T \cup E' \backslash \{e\})_K$ is two-connected, leading to $|\delta(\{z\})|$ new affinely independent vectors lying in the face.

Let $f \in \delta_{G-z}(W_i)$ be an edge incident to w_i. By the last condition, there exists a feasible cycle using f and z. Let T be the set of edges in this cycle that belong to $\bigcup_{i=1}^{p} \delta_{G-z}(W_i)$. The corresponding edges in $\hat{T}$ form a path in $\hat{G}$. Since T was constructed from a feasible cycle, we have $T \cup E' \backslash \{e\} = (T \cup E' \backslash \{e\})_K$.

If $\hat{T}$ is connected, it is a tree and the construction is finished. Otherwise, let W be the set of nodes in the connected component defined by $\hat{T}$. Since $\hat{G}$ is two-connected, there is at least one edge leaving a node $\hat{w}$ different from the node corresponding to W_i. Let f be the corresponding edge in G. Adding f to T increases the size of the subtree $\hat{T}$. It is clear that f is not incident to w_i, and therefore, by the third condition, f belongs to a feasible cycle in $T \cup E' \backslash \{e\}$ and we preserve the fact that $T \cup E' \backslash \{e\} = (T \cup E' \backslash \{e\})_K$. This construction is repeated until $\hat{T}$ forms a tree spanning $\hat{G}$.

We have thus constructed a tree in $\hat{G}$, such that $T \cup E' \backslash \{e\} = (T \cup E' \backslash \{e\})_K$. Again, by Theorem 6.3 in [49], $T \cup E' \backslash \{e\}$ is two-connected and therefore, its incidence vector belongs to $\mathcal{D}om\mathcal{P}_{G,K}$. □

4.8 WEIGHTED PARTITION INEQUALITIES

We now describe inequalities related to a partition of V into subsets $W_1, \ldots, W_p$.

PROPOSITION 4.22 *Let $G = (V, E)$ be a graph, $K > 0$ a given constant, and $W_1, W_2, \ldots, W_p$ $(p \geq 2)$ a partition of V. Then*

$$\frac{1}{2}\sum_{i=1}^{p} \sum_{e \in \delta(W_i)} (K - d_e)x_e \geq (p-1)K \tag{4.17}$$

is a valid inequality for 2CNBR.

PROOF.

Let $(x_e)_{e \in E}$ be variables defining a feasible solution of 2CNBR.

We will first prove the result for $p = 2$, i.e. the partition defines a cut. Note that, by the triangle inequality, an edge $e \in E$ such that $d_e \geq \frac{K}{2}$ does not belong to any cycle of length less than or equal to K, and so $x_e = 0$. This means that

$$d_e x_e \leq \frac{K}{2}\, x_e, \quad e \in E. \tag{4.18}$$

We thus have

$$\begin{aligned}
\frac{1}{2}\sum_{i=1}^{2} \sum_{e\in\delta(W_i)} (K-d_e)x_e \\
&= \sum_{e\in\delta(W_1)} (K-d_e)x_e && \text{since } \delta(W_1)=\delta(W_2), \\
&\geq \sum_{e\in\delta(W_1)} \frac{K}{2}\, x_e && \text{by (4.18)}, \\
&= \frac{K}{2}\, x(\delta(W_1)) \\
&\geq K, && \text{since the solution network must be two-connected.}
\end{aligned}$$

Now suppose that $p > 2$ and that the result holds for all q, $2 \leq q \leq p-1$. Suppose also that there exists a partition $W_1, \ldots, W_p$ of V such that (4.17) is violated, i.e.

$$\frac{1}{2}\sum_{i=1}^{p} \sum_{e\in\delta(W_i)} (K-d_e)x_e < (p-1)K. \tag{4.19}$$

Consider a feasible cycle using at least one edge in $\bigcup_{i=1}^{p} \delta(W_i)$, and denote by C the subset of edges defining this cycle. C is such that

$$\sum_{e\in C} d_e x_e \leq K.$$

Let $\overline{C} = C \cap (\bigcup_{i=1}^{p} \delta(W_i))$. Obviously,

$$\sum_{e\in\overline{C}} d_e x_e \leq K. \tag{4.20}$$

Moreover, the number of elements in $\overline{C}$ is given by $\sum_{e\in\overline{C}} x_e$, implying

$$\sum_{e\in\overline{C}} K x_e = |\overline{C}|K. \tag{4.21}$$

Combining (4.20) and (4.21), we obtain

$$\sum_{e\in\overline{C}}(K-d_e)x_e \geq (|\overline{C}|-1)K. \tag{4.22}$$

Without loss of generality, we may assume that the first $q-1$ subsets of the partition — $W_1,\ldots,W_{q-1}$ — are not incident to $\overline{C}$, while all the others — $W_q, W_{q+1},\ldots,W_p$ — are. We define a new partition of V in q subsets as follows :

$$\overline{W}_i = \begin{cases} W_i & \text{if } i<q \\ \bigcup_{j=q}^{p} W_j & \text{if } i=q \end{cases}$$

This partition is such that

$$\begin{aligned}\frac{1}{2}\sum_{i=1}^{q}\sum_{e\in\delta(\overline{W}_i)}&(K-d_e)x_e \\ &=\frac{1}{2}\sum_{i=1}^{p}\sum_{e\in\delta(W_i)}(K-d_e)x_e-\sum_{e\in T}(K-d_e)x_e,\end{aligned} \tag{4.23}$$

where $T=\left(\bigcup_{i=q}^{p}\delta(W_i)\right)-\delta(\overline{W}_q)$.

By the definition of $\overline{C}$ and $\overline{W}_q$, it is clear that $\overline{C}\subseteq\bigcup_{i=q}^{p}\delta(W_i)$ and $\overline{C}\cap\delta(\overline{W}_q)=\phi$, so $\overline{C}\subseteq T$. Using (4.22), this leads to :

$$\sum_{e\in T}(K-d_e)x_e \geq \sum_{e\in\overline{C}}(K-d_e)x_e \geq (|\overline{C}|-1)K. \tag{4.24}$$

Combining (4.19), (4.23) and (4.24), we have :

$$\begin{aligned}\frac{1}{2}\sum_{i=1}^{q}\sum_{e\in\delta(\overline{W}_i)}&(K-d_e)x_e \\ &<(p-1)K-(|\overline{C}|-1)K=(p-|\overline{C}|)K.\end{aligned}$$

Moreover, $\overline{C}$ is adjacent to at most $|\overline{C}|$ subsets in the original partition, meaning that $p-q+1\leq|\overline{C}|$, or $p-|\overline{C}|\leq q-1$.

We conclude that

$$\frac{1}{2}\sum_{i=1}^{q}\sum_{e\in\delta(\overline{W}_i)}(K-d_e)x_e < (q-1)K$$

which leads to a contradiction since $q < p$, and (4.17) is thus a valid inequality for 2CNBR. □

We call this new class of inequalities *weighted partition inequalities.*

Chapter 5

THE SPECIAL CASE OF RINGS WITH BOUNDED CARDINALITY

An interesting particular case of 2CNBR arises when the number of edges in a ring is bounded, instead of the length of the ring. This corresponds to giving a unit length to each edge. Such a case occurs in the design of self-healing rings.

In this chapter, we derive additional properties for this particular case. In Section 5.1, we compute a lower bound on the number of edges in any feasible solution. This bound is used in Section 5.2 to show that the problem is NP-complete even for K fixed, and is then extended to a new class of valid inequalities in Section 5.3. We end the chapter with additional properties for triangular rings, i.e. the case where $K = 3$.

5.1 A LOWER BOUND ON THE NUMBER OF EDGES

In this section, we compute a lower bound on the number of edges in any feasible solution of 2CNBR. The results of this section are useful for showing that the problem is NP-complete for any fixed $K \geq 3$ and for deriving new valid inequalities.

We first need the following definitions and properties from graph theory. Let $G = (V, E)$ be a connected graph with $n = |V|$ nodes and $m = |E|$ edges, and let T be a spanning tree in G. The addition of an edge to T creates exactly one cycle, called a *fundamental cycle*. Since there are $m - n + 1$ chords in G, there are exactly $m - n + 1$ fundamental cycles associated with each spanning tree.

Now, suppose the edges of G are ordered as $e_1, \ldots, e_m$, and we have a set of cycles σ_i, $i = 1, \ldots, k$. Then, we can define a *cycle matrix* $C = (c_{ij})_{k \times m}$ in which

$$c_{ij} = \begin{cases} 1 & \text{if } \sigma_i \text{ contains edge } e_j, \\ 0 & \text{otherwise.} \end{cases}$$

If the cycles σ_i are the fundamental cycles associated to a spanning tree T, this matrix is called *fundamental cycle matrix*. Moreover, if edges in G are numbered starting from the edges not belonging to T, and the fundamental cycles are numbered accordingly, then the fundamental cycle matrix has the following form :

$$C = (I_\mu; C_t),$$

where I_μ is an identity matrix of dimension $\mu = m - n + 1$, and C_t is the remaining matrix of dimension $\mu \times (n - 1)$ corresponding to the branches of T.

Figure 5.1 shows an example of graph with its associated fundamental cycle matrix, where the fundamental cycles are defined by the spanning tree drawn in bold.

It is clear that the rank of the fundamental cycle matrix C is $\mu = m - n + 1$. With respect to scalars in $\{0, 1\}$ and the addition modulo 2, the fundamental cycles are independent, and we can define a vector space for which the fundamental cycles form a basis. This vectorial space is called the *cycle space*, and all the other cycles in G can be obtained as a linear combination of rows representing the fundamental

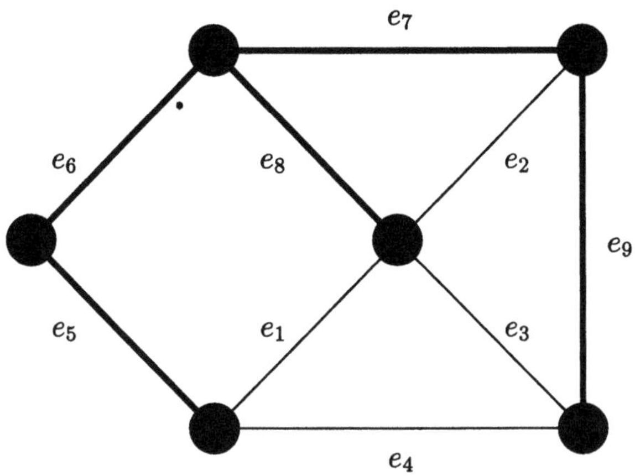

$$C = \left(\begin{array}{cccc|ccccc} e_1 & e_2 & e_3 & e_4 & e_5 & e_6 & e_7 & e_8 & e_9 \\ \hline 1 & 0 & 0 & 0 & 1 & 1 & 0 & 1 & 0 \\ 0 & 1 & 0 & 0 & 0 & 0 & 1 & 1 & 0 \\ 0 & 0 & 1 & 0 & 1 & 1 & 1 & 0 & 1 \\ 0 & 0 & 0 & 1 & 1 & 1 & 1 & 0 & 1 \end{array} \right)$$

Figure 5.1. An example of graph and its fundamental cycle matrix

cycles in C. The dimension μ of the cycle space is also called the *cyclomatic number* of the graph.

More details about fundamental cycles and the cycle space can be found e.g. in Berge [8].

We can now prove the first important result of this section.

THEOREM 5.1 *Let $G = (V, E)$ be a two-*edge-*connected network with $n = |V|$ nodes and $m = |E|$ edges. If there exists a covering of the network by cycles using at most K nodes, then there exists such a covering using at most $\mu = m - n + 1$ independent cycles.*

PROOF.

Since G is two-edge-connected, each edge belongs to a cycle, which can be obtained by a linear combination of the elements of a basis of the cycle space. It means that each edge must belong to at least one cycle in a basis, or, in other words, that a basis of the cycle space covers the edges of the network. Moreover, each cycle matrix of rank μ defines a basis of the cycle space. It means that the set of cycles defined by any cycle matrix of rank μ — and in particular, fundamental cycles — covers the edges of the network.

Let C be a fundamental cycle matrix for G. If a subset of the fundamental cycles that use at most K nodes covers the edges, it forms the requested covering. Otherwise, we transform C in a cycle matrix C' of rank μ such that the subset of cycles of C' that use at most K nodes covers the network.

In order to obtain that transformation, we now show that a cycle matrix of rank μ, defining r cycles using at most K nodes, and such that these r cycles do not cover the set of edges E, can be transformed into a cycle matrix of rank μ, defining $r + 1$ cycles using at most K nodes. By applying this construction iteratively to the fundamental cycle matrix C, we create a sequence of matrices of rank μ, thus covering the edges of the network by independent cycles, and such that the number of cycles using at most K nodes is increased by one at each step. We stop when these cycles using at most K nodes cover the network, i.e. these form the requested covering. This occurs in a finite number of steps, since in the worst case, we end with a matrix containing μ cycles using at most K nodes, and that cover the network.

Let C_1 be a $\mu \times m$ cycle matrix of rank μ and suppose there are r cycles using at most K nodes in C_1. Without loss of generality, we can suppose these cycles form the r first rows of C_1, i.e. C_1 is of the form

$$C_1 = \begin{pmatrix} C_r \\ C_t \end{pmatrix},$$

where C_r is a $r \times m$ cycle matrix such that each cycle defined by C_r uses at most K nodes and C_t is a $(\mu - r) \times m$ cycle matrix such that each cycle defined by C_t uses at least $K + 1$ nodes.

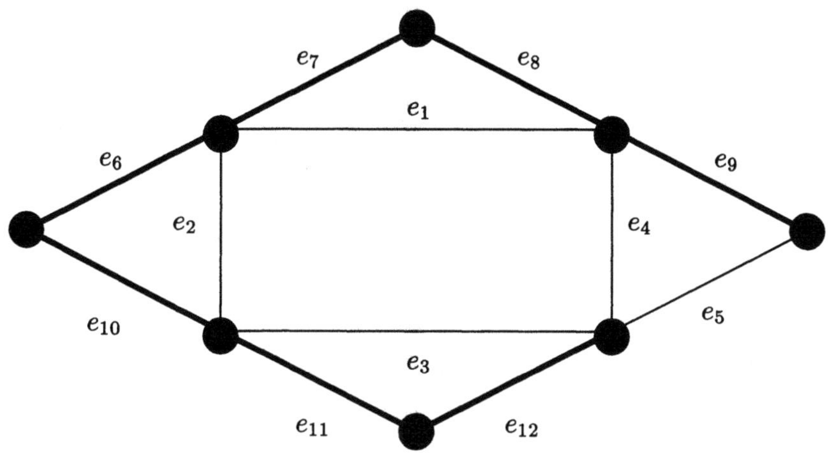

Figure 5.2. An illustration of Theorem 5.1

Suppose cycles in C_r do not cover the network, i.e. there exists an edge $e \in E$ which does not belong to any cycle in C_r. This edge must belong to a cycle σ using at most K nodes. Since C_1 is of rank μ, σ can be obtained by a linear combination of its rows. Moreover, since no cycle in C_r uses e, the linear combination uses at least one row of C_t. Replacing this row of C_t by the linear combination of rows defining σ, we obtain a new cycle matrix C_2 of same rank μ and defining $r + 1$ cycles using at most K nodes. □

Note that the cycles in the cover do not necessarily correspond to a spanning tree. The constructive proof of Theorem 5.1 is illustrated by the following example. Consider the graph of Figure 5.2 with the fundamental cycles associated to the spanning tree drawn in bold, and suppose the rings must use at most 3 nodes. The fundamental cycle matrix is

$$C = \left(\begin{array}{ccccc|ccccccc|c}
e_1 & e_2 & e_3 & e_4 & e_5 & e_6 & e_7 & e_8 & e_9 & e_{10} & e_{11} & e_{12} & \\
\hline
1 & 0 & 0 & 0 & 0 & 0 & 1 & 1 & 0 & 0 & 0 & 0 & \sigma_1 \\
0 & 1 & 0 & 0 & 0 & 1 & 0 & 0 & 0 & 1 & 0 & 0 & \sigma_2 \\
0 & 0 & 1 & 0 & 0 & 0 & 0 & 0 & 0 & 0 & 1 & 1 & \sigma_3 \\
\hline
0 & 0 & 0 & 1 & 0 & 1 & 1 & 1 & 0 & 1 & 1 & 1 & \sigma_4 \\
0 & 0 & 0 & 0 & 1 & 1 & 1 & 1 & 1 & 1 & 1 & 1 & \sigma_5
\end{array} \right).$$

The matrix C_r of cycles using at most 3 nodes consists of the first 3 rows of C. However, the column corresponding to e_4 in C_r is null, meaning e_4 is not covered by C_r. But $e_4 \in \overline{\sigma} = \{e_4, e_5, e_9\}$ and $\overline{\sigma}$ is obtained by combining σ_4 and σ_5. We can thus replace σ_4 by $\overline{\sigma}$ and we obtain the new matrix

$$C' = \left(\begin{array}{ccccc|ccccccc|c}
e_1 & e_2 & e_3 & e_4 & e_5 & e_6 & e_7 & e_8 & e_9 & e_{10} & e_{11} & e_{12} & \\
\hline
1 & 0 & 0 & 0 & 0 & 0 & 1 & 1 & 0 & 0 & 0 & 0 & \sigma_1 \\
0 & 1 & 0 & 0 & 0 & 1 & 0 & 0 & 0 & 1 & 0 & 0 & \sigma_2 \\
0 & 0 & 1 & 0 & 0 & 0 & 0 & 0 & 0 & 0 & 1 & 1 & \sigma_3 \\
0 & 0 & 0 & 1 & 1 & 0 & 0 & 0 & 1 & 0 & 0 & 0 & \overline{\sigma} \\
\hline
0 & 0 & 0 & 0 & 1 & 1 & 1 & 1 & 1 & 1 & 1 & 1 & \sigma_5
\end{array} \right).$$

Note that cycles using at most 3 nodes are given by the first 4 rows, and no column of the corresponding sub-matrix is null, meaning $\sigma_1, \sigma_2, \sigma_3$ and $\overline{\sigma}$ form a covering of the network as requested in Theorem 5.1.

Theorem 5.1 is the key to establish the main result of this section.

THEOREM 5.2 *Let $G = (V, E)$ be a two-connected network with $n = |V|$ nodes and $m = |E|$ edges, such that there exists a covering of the network by cycles using at most K nodes. Then,*

$$m \geq M(n, K) := n + \min\left(\left\lceil \frac{n-K}{K-2} \right\rceil, \left\lceil \frac{n}{K-1} \right\rceil\right), \tag{5.1}$$

i.e. G contains at least $M(n, K)$ edges.

PROOF.

From Theorem 5.1, there exists a covering of G by at most $m-n+1$ independent cycles using at most K nodes. We consider two disjoint cases :

1. *There exists a covering satisfying Theorem 5.1 and using at most $m-n$ cycles.*

   In this case, the sum of the number of edges used in each cycle is greater than or equal to m, since the cycles cover the network. Moreover, since each cycle uses at most K edges, this sum is less than or equal to $(m-n)K$ implying

   $$m \leq (m-n)K,$$

   which can be written

   $$m \geq n + \frac{n}{K-1}.$$

   Since m is integral, this implies that

   $$m \geq n + \left\lceil \frac{n}{K-1} \right\rceil . \tag{5.2}$$

2. *All coverings satisfying Theorem 5.1 use exactly $m-n+1$ cycles.*

   It means that such a covering is also a basis of the cycle space. We first show that each cycle in the covering shares at least one edge with the others. Suppose it is not the case for some cycle σ in the covering. Since the network is two-connected, the subset of nodes of the cycle adjacent to nodes that do not belong to the cycle (the *boundary* of the cycle) contains at least two different nodes — otherwise there exists an articulation point. Let u and v be two nodes in the boundary, as depicted in Figure 5.3. Using again the fact that the network is two-connected, we know that there exists a path P, outside σ, between u and v. Two other paths P_1 and P_2 between u and v form σ. Combining P with P_1 and P_2 respectively, we obtain two new cycles σ_1 and σ_2. Since the covering defines a basis, σ_1 and σ_2 can be obtained by a linear combination of the cycles in the covering.

   If we suppose the edges are ordered starting from those in P_1 then P_2, and finally the remaining edges, the vector corresponding to σ has the form

   $$(\underbrace{1\ldots1}_{P_1}\ \underbrace{1\ldots1}_{P_2}\ 0\ldots0)$$

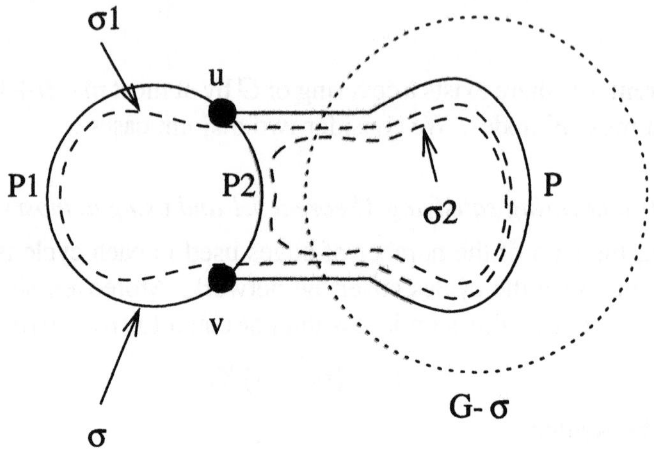

Figure 5.3. Combinations of a cycle σ and an outside path P to generate cycles σ_1 and σ_2

while the other cycles in the base have the form

$$(\underbrace{0 \ldots 0}_{P_1} \; \underbrace{0 \ldots 0}_{P_2} \times \ldots \times)$$

since we supposed σ had no edge in common with any other cycle in the covering. Moreover, the vector corresponding to σ_1 has the form

$$(\underbrace{1 \ldots 1}_{P_1} \; \underbrace{0 \ldots 0}_{P_2} \times \ldots \times).$$

It is clear from the form of those vectors that σ_1 cannot be obtained by a linear combination of cycles in the covering, which leads to a contradiction.
So each cycle in the basis shares at least one edge with the others. The sum of the number of edges used in each cycle is thus greater than or equal to $2m - n$, since at least $m - n$ repetitions of edge occur in this counting. This sum is also less than or equal to $(m - n + 1)K$ since each cycle uses at most K edges. We can conclude that

$$2m - n \leq (m - n + 1)K,$$

or, since m is integral,

$$m \geq n + \left\lceil \frac{n - K}{K - 2} \right\rceil . \tag{5.3}$$

Since one of these two cases must occur, (5.2) or (5.3) is satisfied, thus the number of edges m satisfies (5.1). □

Theorem 5.2 provides a lower bound on the number of edges in a feasible solution. This bound is tight, as we will see in the proof of Theorem 5.6, and it is useful for showing that the problem is NP-complete, which is done in the next section. Moreover, a similar application of Theorem 5.1 leads to some classes of strong valid inequalities that are described in Section 5.3.

5.2 COMPLEXITY

In this section, we show that the recognition version of the 2CNBR problem is NP-complete for any fixed value of the bound K.

PROBLEM 5.3 (R2CNBR) *Let $G = (V, E)$ be a graph, $K \geq 3$ a given constant and $B \geq 0$ a real number. To each edge $e \in E$ is associated a cost c_e and a unit length $d_e = 1$. Does there exists a subset $F \subseteq E$ of edges such that F_K is two-connected and $c(F) \leq B$?*

THEOREM 5.4 *R2CNBR is NP-complete for any $K \geq 3$.*

PROOF.

It is easy to see that R2CNBR belongs to NP. We show that the Hamiltonian cycle problem reduces to R2CNBR, for any fixed $K \geq 3$. Let $G = (V, E)$ be a graph with $n = |V|$ nodes and $m = |E|$ edges, and suppose $V = \{v_1, \ldots, v_n\}$. The Hamiltonian cycle problem consists in determining if there exists a Hamiltonian cycle in G.

This problem is NP-complete and it can be transformed into R2CNBR in the following way.

If $K \geq n$, finding a Hamiltonian cycle is equivalent to finding a two-connected network of cost less than or equal to n, for the same graph, with unit edge costs. But $K \geq n$ implies that R2CNBR is equivalent to the two-connected network problem, since the constraints on the rings are redundant. Thus the Hamiltonian cycle problem reduces to R2CNBR if $K \geq n$.

If $K < n$, define a graph $G' = (V', E')$ with

$$V' = \bigcup_{k=0}^{K-2} V^k, \text{ where } V^k = \{v_1^k, \dots, v_n^k\}, \quad k = 0, \dots, K-2,$$

and

$$\begin{aligned} E' \quad = \quad & \bigcup_{i,j:\{v_i,v_j\} \in E} \{\{v_i^0, v_j^0\}, \{v_i^0, v_j^{K-2}\}, \{v_j^0, v_i^{K-2}\}\} \\ & \cup \bigcup_{k=0}^{K-3} \bigcup_{i=0}^{n} \{\{v_i^k, v_i^{k+1}\}\} \end{aligned}$$

A unit cost is again assigned to each edge. This transformation is illustrated in Figure 5.4 for $K = 5$.

G' is composed of $n' = (K-1)n$ nodes and $m' = 3m + (K-2)n$ edges. The transformation can thus be performed in polynomial time and space. We will show that there exists a feasible solution of 2CNBR of cost less than or equal to Kn in G' if and only if there exists a Hamiltonian cycle in G, and conversely.

Suppose there exists a solution F of 2CNBR of cost less than or equal to Kn in G'. Since edge costs are unitary, the number of edges in F is equal to the cost of the solution, $c(F)$. By Theorem 5.2, we thus have

$$c(F) \geq n' + \min\left(\left\lceil \frac{n' - K}{K-2} \right\rceil, \left\lceil \frac{n'}{K-1} \right\rceil\right).$$

But $n' = (K-1)n$, so we have

$$\frac{n'}{K-1} = n$$

and

$$\frac{n' - K}{K-2} = \frac{(K-1)n - K}{K-2} = n + \frac{n-K}{K-2} > n = \frac{n'}{K-1},$$

since $n > K$ and $K \geq 3$. Thus,

$$Kn \geq c(F) \geq n' + \left\lceil \frac{n'}{K-1} \right\rceil = (K-1)n + n = Kn,$$

and F contains exactly Kn edges.

Since F is two-connected, the degree of each node in this solution is at least two. Moreover, for each $v_i^k, i = 1, \ldots, n,\ k = 1, \ldots, K-3$, there are only two possible edges, and these edges must belong to any feasible solution. There are $(K-2)n$ such edges that form n paths of length $K-2$, $(v_i^0, v_i^1, \ldots, v_i^{K-2})$, as illustrated in bold in Figure 5.5. Since each edge in these paths must belong to a cycle using at most K nodes, two edges (v_i^{K-2}, w) and (w, v_i^0) must be in the solution to close each path and form a feasible cycle. It is clear that the only possible choice for w is a node $v_j^0 \in V^0$.

In this way, we get a subgraph (V, F) with Kn edges, with exactly n edges in $F(V^0)$, corresponding to edges in E. We must show that these edges form a Hamiltonian cycle in G. Since there are exactly n edges, it is sufficient to show that these induce a two-connected subgraph of G, i.e. that $F(V^0)$ is two-connected. Let v_i^0 and v_j^0 be two nodes in V^0. In F, there are two node-disjoint paths from v_i^0 to v_j^0, since F is two-connected. If these paths use only nodes in V^0, the expected result holds. Otherwise, each node not belonging to V^0 must appear in a sequence of the form $v_k^0, v_k^1, \ldots, v_k^{K-2}, v_l^0$. By our previous construction, this imply that (v_k^0, v_l^0) is also in the solution. Replacing each sequence of the form $v_k^0, v_k^1, \ldots, v_k^{K-2}, v_l^0$ using nodes outside V^0 by the corresponding shortcut (v_k^0, v_l^0) leads to two disjoint paths from v_i^0 to v_j^0 in $E(V^0)$.

So $E(V^0)$ is two-connected and the n edges in $F(V^0)$ correspond to a Hamiltonian cycle in G.

Conversely, it is easy to see that we can complete an edge set $F(V^0)$ corresponding to a Hamiltonian cycle in G, defined by the sequence of nodes $v_{i_1}, v_{i_2}, \ldots, v_{i_n}$, $v_{i_{n+1}} = v_{i_1}$, into a feasible solution F of 2CNBR with Kn edges by adding the paths $v_{i_k}^0, v_{i_k}^1, \ldots, v_{i_k}^{K-2}, v_{i_{k+1}}^0$ for $k = 1, \ldots, n$ to the edges already in $F(V^0)$.

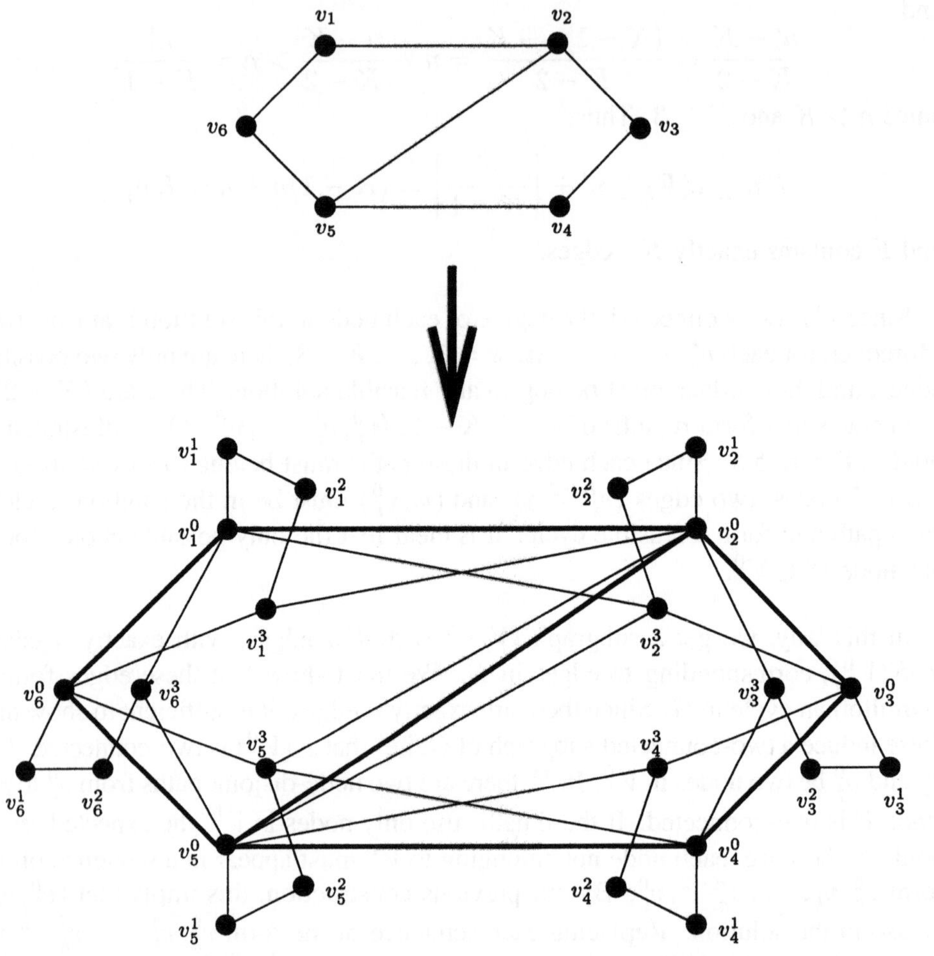

Figure 5.4. Transformation of Hamiltonian cycle into 2CNBR

This concludes the proof of Theorem 5.4. □

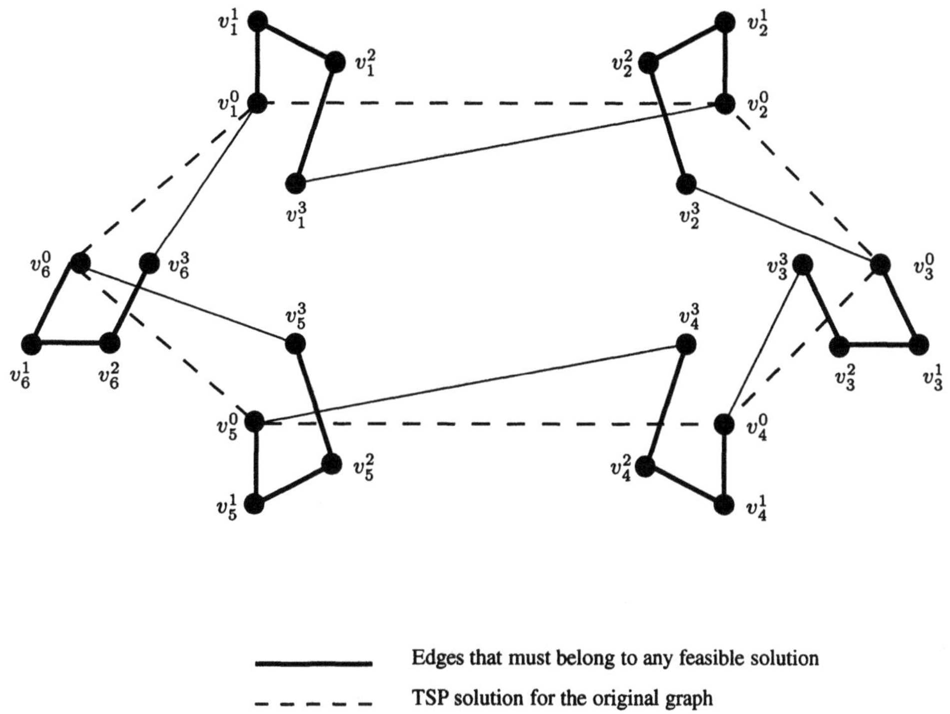

Figure 5.5. A solution of R2CNBR and the corresponding Hamiltonian cycle

5.3 CYCLOMATIC INEQUALITIES

From the lower bound on the number of edges in a feasible solution obtained in Section 5.1, we can conclude that

$$\sum_{e \in E} x_e \geq M(n,K) = n + \min\left(\left\lceil \frac{n-K}{K-2} \right\rceil, \left\lceil \frac{n}{K-1} \right\rceil\right)$$

is a valid inequality for 2CNBR. In this section, we extend this result to new classes of valid inequalities for 2CNBR.

Our first result is an extension of Theorem 5.2 to a partition of V.

PROPOSITION 5.5 *Let* $G = (V, E)$ *be a graph with* $n = |V|$ *nodes,* $K \geq 3$ *a given constant, and* $W_1, W_2, \ldots, W_p$ $(p \geq 2)$ *a partition of* V*. Then*

$$\frac{1}{2}\sum_{i=1}^{p} x(\delta(W_i)) \geq M(p, K) \tag{5.4}$$

is a valid inequality for 2CNBR.

PROOF.

Let F be a feasible solution to the 2CNBR problem and let $\hat{G}$ denote the contracted graph $(V, F)/W_1/\ldots/W_p$, and $\hat{m}$ the number of edges in $\hat{G}$. We show that $\hat{m} \geq M(p, K)$.

It is easy to see that $\hat{G}$ is two-edge-connected and that each edge in $\hat{G}$ belongs to a cycle using at most K edges.

If $\hat{G}$ is two-connected, Theorem 5.2 holds for $\hat{G}$, i.e $\hat{m} \geq M(p, k)$. Otherwise, $\hat{G}$ contains $q \geq 1$ articulation points. Let z be one of these nodes, obtained after the contraction of a subset $Z \subseteq V$ of nodes in G (Figure 5.6(a,b)). Since F is two-connected, the boundary of Z contains at least two nodes. Let u be one of these nodes. Replacing z by two nodes z_1 and z_2, and replacing each edge incident to z by an edge connected to z_1 if the corresponding edge in G was connected to u, and by an edge connected to z_2 otherwise, we obtain a graph with one articulation point less, as illustrated in Figure 5.6(c). However, it is possible that this new graph contains some edges not belonging to a cycle using at most K edges. In this case, let e be one of these edges. Before replacing z, e belonged to at least one feasible cycle. Each of these cycles has been replaced by a path from z_1 to z_2. It is easy to see that one of these paths must contain at most $K-1$ edges, otherwise the corresponding edge in F cannot belong to a feasible cycle. Therefore, adding an edge linking z_1 and z_2 is sufficient to create a feasible cycle containing e (Figure 5.6(d)). Repeating this construction for each articulation point, we obtain a two-connected graph $\overline{G}$ such that each edge in $\overline{G}$ belongs to a cycle using at most K edges. Since $\overline{G}$ has $p+q$ nodes, it contains at least $M(p+q, K)$ edges by Theorem 5.2, and by our construction, it contains at most $\hat{m}+q$ edges. Therefore,

$$\hat{m} \geq M(p+q, K) - q = p + \min\left(\left\lceil \frac{p+q-K}{K-2} \right\rceil, \left\lceil \frac{p+q}{K-1} \right\rceil\right) \geq M(p, K),$$

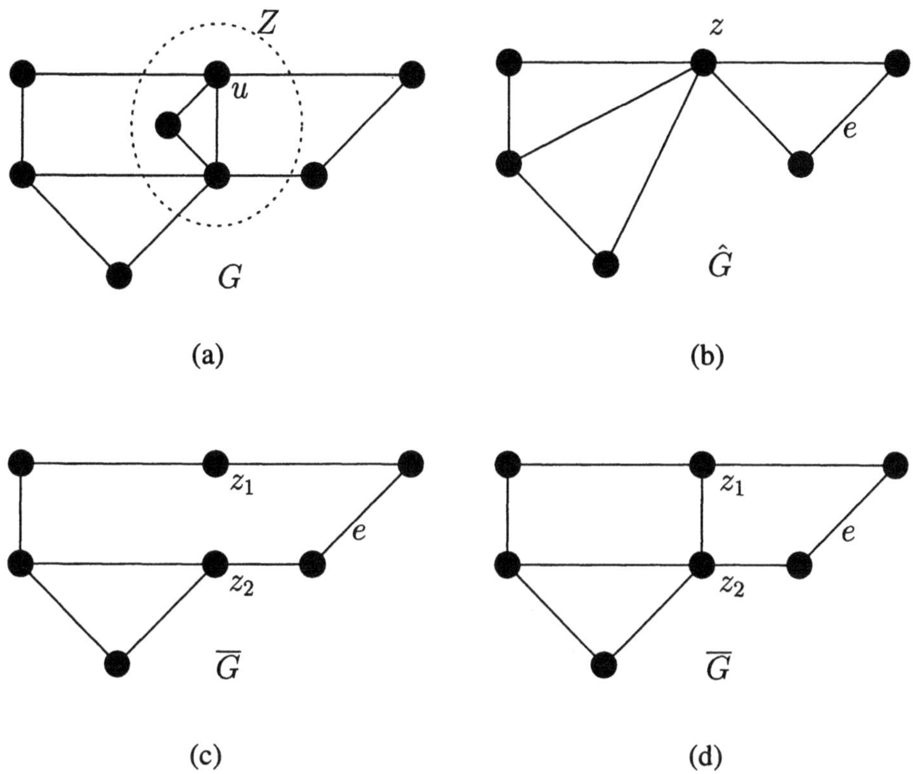

Figure 5.6. A graph G, the contracted graph $\hat{G}$ and the two steps of $\overline{G}$ construction, with a bound $K = 4$.

and we have proved Proposition 5.5. □

Inequalities (5.4) are called *cyclomatic inequalities*. The next theorem shows that the inequality bounding the total number of edges (i.e. $p = n$) is facet-defining for complete graphs.

THEOREM 5.6 *Let $G = (V, E)$ be a complete graph with $V = \{v_1, \ldots, v_n\}$, and $3 \leq K < n$ a given constant. Then $x(E) \geq M(n, K)$ defines a facet of $\mathcal{D}om\mathcal{P}_{G,K}$.*

PROOF.

We consider separately the two cases corresponding to the two possible values of $M(n, K)$.

$\mathbf{M(n,K) = n + \left\lceil \frac{n-K}{K-2} \right\rceil}$.

Let $b^T x \geq \beta$ be a facet-defining inequality such that the face induced by $x(E) \geq M(n,k)$ in $\mathcal{D}om\mathcal{P}_{G,K}$ is contained in the facet F_b induced by $b^T x \geq \beta$. Our aim is to show that b_e has the same value for all $e \in E$.

Consider a permutation $\{i_1, \ldots, i_n\}$ of $\{1, \ldots, n\}$, and the two following sets of edges :

$$
\begin{aligned}
F_1 &= \bigcup_{j=1}^{n-1} \{\{v_{i_j}, v_{i_{j+1}}\}\} \cup \{\{v_{i_1}, v_{i_n}\}\} \cup \{\{v_{i_1}, v_{i_K}\}\} \\
&\cup \bigcup_{j=1}^{\left\lceil \frac{n-2}{2(K-2)} \right\rceil - 1} \{\{v_{i_{K+(j-1)(K-2)}}, v_{i_{n+1-j(K-2)}}\}\} \\
&\cup \bigcup_{j=1}^{\left\lceil \frac{n-K}{2(K-2)} \right\rceil - 1} \{\{v_{i_{n+1-j(K-2)}}, v_{i_{K+j(K-2)}}\}\}, \\
F_2 &= \bigcup_{j=2}^{n-1} \{\{v_{i_j}, v_{i_{j+1}}\}\} \cup \{\{v_{i_1}, v_{i_n}\}\} \\
&\quad \cup \{\{v_{i_1}, v_{i_K}\}\} \cup \{\{v_{i_2}, v_{i_{K+1}}\}\} \\
&\cup \bigcup_{j=1}^{\left\lceil \frac{n-2}{2(K-2)} \right\rceil - 1} \{\{v_{i_{K+(j-1)(K-2)}}, v_{i_{n+1-j(K-2)}}\}\} \\
&\cup \bigcup_{j=1}^{\left\lceil \frac{n-K}{2(K-2)} \right\rceil - 1} \{\{v_{i_{n+1-j(K-2)}}, v_{i_{K+j(K-2)}}\}\}.
\end{aligned}
$$

These constructions are illustrated in Figure 5.7 for $n = 12$ and $K = 4$. It is easy to see that F_1 and F_2 define feasible solutions of 2CNBR, and that

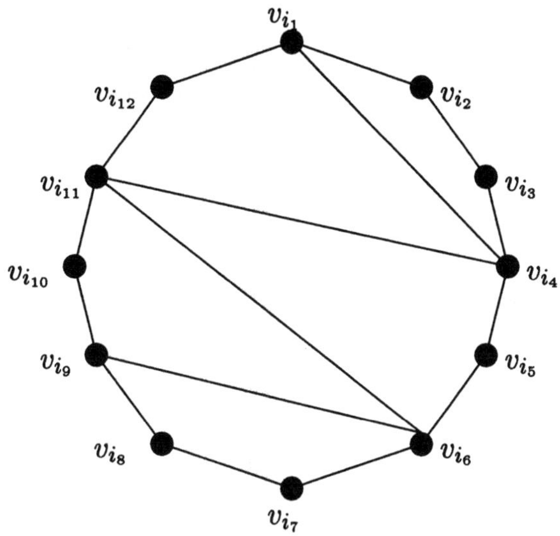

F_1

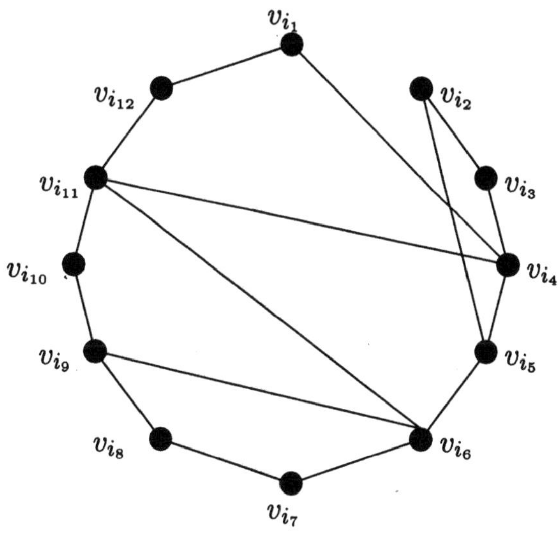

F_2

Figure 5.7. F_1 and F_2

$|F_1| = |F_2| = M(n, K)$. Therefore, the incidence vectors of F_1 and F_2 lie in $\mathcal{F}_b$. Since F_2 is obtained from F_1 by replacing edge $\{v_{i_1}, v_{i_2}\}$ by $\{v_{i_2}, v_{i_{K+1}}\}$, we obtain that $b_{\{v_{i_1},v_{i_2}\}} = b_{\{v_{i_2},v_{i_{K+1}}\}}$.

Consider three distinct nodes v_i, v_j and v_k. Since the permutation defined above was arbitrary, any permutation such that $i_1 = i, i_2 = j$ and $i_{K+1} = k$ leads to $b_{\{v_i,v_j\}} = b_{\{v_j,v_k\}}$. If v_i, v_j, v_k and v_l are four distinct nodes, we obtain by transitivity that $b_{\{v_i,v_j\}} = b_{\{v_j,v_k\}} = b_{\{v_k,v_l\}}$, and therefore b_e has the same value for all $e \in E$.

$\mathbf{M(n, K) = n + \left\lceil \frac{n}{K-1} \right\rceil}$.

Let $b^T x \geq \beta$ be a facet-defining inequality such that the face induced by $x(E) \geq M(n, k)$ in $\mathcal{D}om\mathcal{P}_{G,K}$ is contained in the facet $\mathcal{F}_b$ induced by $b^T x \geq \beta$. Our aim is to show that b_e has the same value for all $e \in E$.

Consider a permutation $\{i_1, \ldots, i_n\}$ of $\{1, \ldots, n\}$, and the two following sets of edges :

$$\begin{aligned}
F_3 &= \{\{v_{i_1}, v_{i_n}\}\} \cup \{\{v_{i_1}, v_{i_{n-K+2}}\}\} \cup \bigcup_{j=1}^{n-1} \{\{v_{i_j}, v_{i_{j+1}}\}\} \\
&\quad \cup \bigcup_{j=1}^{\lceil \frac{n}{K-1} \rceil - 1} \{\{v_{i_{1+(j-1)(K-1)}}, v_{i_{1+j(K-1)}}\}\}, \\
F_4 &= \{\{v_{i_2}, v_{i_n}\}\} \cup \{\{v_{i_2}, v_{i_{n-K+2}}\}\} \cup \bigcup_{j=1}^{n-1} \{\{v_{i_j}, v_{i_{j+1}}\}\} \\
&\quad \cup \bigcup_{j=1}^{\lceil \frac{n}{K-1} \rceil - 1} \{\{v_{i_{1+(j-1)(K-1)}}, v_{i_{1+j(K-1)}}\}\}.
\end{aligned}$$

These constructions are illustrated in Figure 5.8 for $n = 15$ and $K = 4$. It is easy to see that F_3 and F_4 define feasible solutions of 2CNBR, and that $|F_3| = |F_4| = M(n, K)$. Therefore, the incidence vectors of F_3 and F_4 lie in $\mathcal{F}_b$. Since F_4 is obtained from F_3 by replacing edges $\{v_{i_1}, v_{i_n}\}$ and $\{\{v_{i_1}, v_{i_{n-K+2}}\}\}$ by $\{v_{i_2}, v_{i_n}\}$ and $\{\{v_{i_2}, v_{i_{n-K+2}}\}\}$, we obtain that $b_{\{v_{i_1},v_{i_n}\}} + b_{\{v_{i_1},v_{i_{n-K+2}}\}} = b_{\{v_{i_2},v_{i_n}\}} + b_{\{v_{i_2},v_{i_{n-K+2}}\}}$.

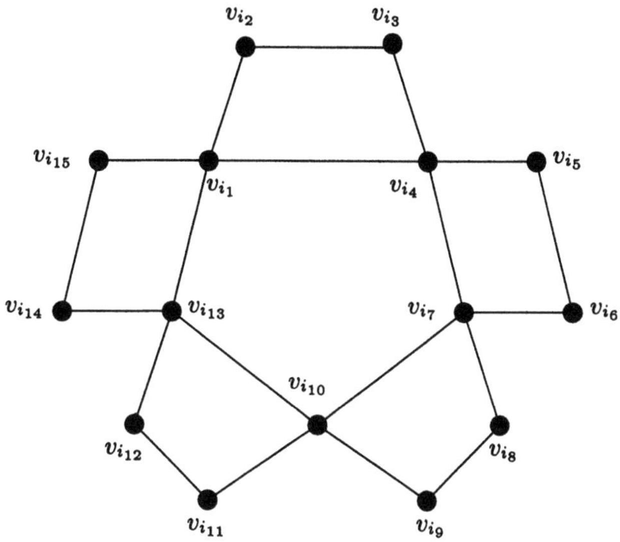

F_3

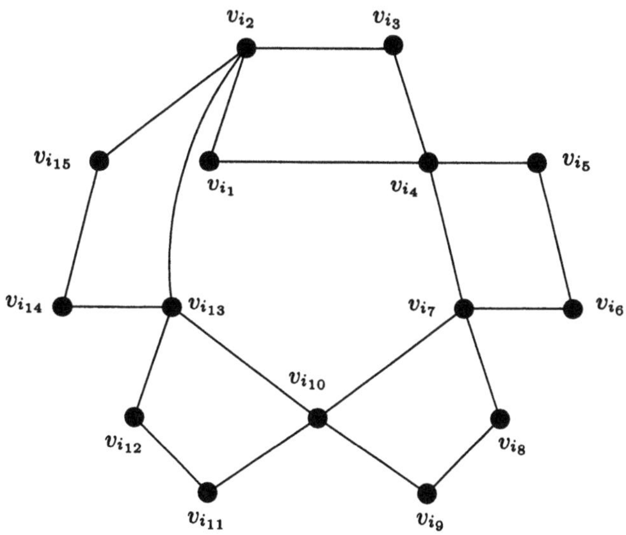

F_4

Figure 5.8. F_3 and F_4

Consider four distinct nodes v_i, v_j, v_k and v_l. Since the permutation defined above was arbitrary, any permutation such that $i_1 = i, i_2 = j, i_n = k$ and $i_{n-K+2} = l$ leads to

$$b_{\{v_i,v_k\}} + b_{\{v_i,v_l\}} = b_{\{v_j,v_k\}} + b_{\{v_j,v_l\}}.$$

Similarly, any permutation such that $i_1 = k, i_2 = l, i_n = i$ and $i_{n-K+2} = j$ leads to

$$b_{\{v_i,v_k\}} + b_{\{v_j,v_k\}} = b_{\{v_i,v_l\}} + b_{\{v_j,v_l\}}.$$

This leads to $b_{\{v_i,v_k\}} = b_{\{v_j,v_l\}}$.

It remains to be shown that two edges having a common endpoint have the same coefficient. If v_i, v_j, v_k, v_l and v_m are five distinct nodes, we obtain by transitivity that $b_{\{v_i,v_j\}} = b_{\{v_l,v_m\}} = b_{\{v_i,v_k\}}$, and therefore b_e has the same value for all $e \in E$.

□

5.4 TRIANGULAR RINGS

We now examine the case where rings are triangles, i.e. $K = 3$. In this particular case, ring-cut inequalities studied in Section 4.4 take a particular form and we show in this section that they are always facet-defining.

The following proposition characterizes the ring-cut graph for triangular rings.

PROPOSITION 5.7 *Let $G = (V, E)$ be a graph and $W \subseteq V$ a subset of nodes, $\phi \neq W \neq V$, such that $2 \leq |W| \leq |V| - 2$. Then $RCE_{W,3} = \phi$.*

PROOF.

Let $e, f \in \delta(W)$ be two edges in the cut and suppose there is an edge between the corresponding nodes in the ring-cut graph. In other words, $(C_{e,f})_K$ is two-connected, and e and f belong to a feasible ring. Since feasible rings are triangles, e and f must have a common endpoint z. Since both W and $V \backslash W$ contain at least two nodes, z is an articulation point in $(C_{e,f})_K$, which leads to a contradiction. □

Since the ring-cut graph corresponding to triangular rings is empty, the entire cut $\delta(W)$ defines an independent subset in the ring-cut graph, and therefore the corresponding inequality is $x(\delta(W)) \geq 3$.

COROLLARY 5.8 *Let* $G = (V, E)$ *be a complete graph and* $W \subseteq V$ *a subset of nodes,* $\phi \neq W \neq V$*, such that* $2 \leq |W| \leq |V| - 2$*. Then* $x(\delta(W)) \geq 3$ *defines a facet of* $\mathcal{D}om\mathcal{P}_{G,K}$.

PROOF.

Follows from Proposition 5.7 and Theorem 4.18. □

Chapter 6

A BRANCH-AND-CUT ALGORITHM

In this chapter, we describe the details of the implementation of a Branch-and-Cut algorithm for the 2CNBR problem. In Section 6.1, we address the problem of determining if a network is feasible. In Sections 6.2 to 6.8, we study the separation problems for all the valid inequalities described in the previous chapters. The order in which inequalities are presented here is not the same as in the previous chapters, because we wanted to group inequalities having similar separation procedures. Section 6.9 outlines the Branch-and-Cut procedure and explains some strategy and parameters choices. The last section of this chapter presents a summary of the polyhedral results obtained in this work.

6.1 CHECKING FEASIBILITY

An important issue for implementing efficient algorithms for the 2CNBR problem is feasibility checking. A simple (but inefficient) way to evaluate whether a graph G is feasible is to check if it is two-connected, and if each edge belongs to a feasible ring (by computing a shortest path not using the edge, between its endpoints). In other words, it is the same as checking that $G = G_K$ and that G_K is two-connected.

As described in Section 4.1, the second condition imposes that the incidence vector of G belongs to $\mathcal{D}om\mathcal{P}_{G,K}$. In itself, checking if this condition is satisfied is an important issue, e.g. in the separation of ring-cut and subset inequalities. Moreover, we are interested in feasible graphs of minimum cost. If $G \neq G_K$, the cost of G_K is less than the cost of G. Therefore, if G_K is two-connected, it provides a feasible solution of cost less than the cost of the (infeasible) solution G.

Instead of checking feasibility of G, we are thus more interested in the feasibility of G_K, which is simply a two-connectivity test for G_K. This test can be performed by an $O(m)$ depth-first search algorithm (see e.g. Aho et al. [3]). The bottleneck in this approach is that we still have to construct G_K, which requires to check for each edge if it belongs to a feasible ring. Instead of doing this check before testing two-connectivity, we propose to adapt the depth-first search algorithm in order to reduce the number of searches for feasible cycles.

First, we recall some basic concepts about depth-first search algorithms. A depth-first search of an undirected graph $G = (V, E)$ is a method aiming at visiting the nodes of the graph in the following manner. We select and visit a starting node v (the *root* of the depth-first search). Then we select any edge $\{v, w\}$ incident to v, and visit w. In general, suppose x is the most recently visited node. The search is continued by selecting some unexplored edge $\{x, y\}$ incident to x. If y has been previously visited, we find another new edge incident to x. If y has not been previously visited, then we visit y and begin the search anew starting at node y. After completing the search through all paths beginning at y, the search returns to x, the node from which y was first reached. The process of selecting unexplored edges incident to x is continued until the list of these edges is exhausted.

Algorithm 6.1 Generic depth-first search of an undirected graph $G = (V, E)$

Data : a graph $G = (V, E)$, a node $v \in V$.
Notation :
$d(i)$: degree of i in G.
$\Gamma_j(i)$: j-th node adjacent to i in G.

```
for all i ∈ V do
   p(i) := −1; num(i) := ∞
p(v) := v; num(v) := 1; i := v; j(i) := 1; last_num := 1

while (j(i) ≤ d(i)) or (i ≠ v) do
   if j(i) = d(i) then {all the edges leaving i have been examined}
      i := p(i) {backtracking}
   else {we examine the next edge leaving i}
      if p(Γ_{j(i)}(i)) = −1 then {tree edge}
         add {i, Γ_{j(i)}(i)} to T
         p(Γ_{j(i)}(i)) := i
         i := Γ_{j(i)}(i)
         last_num := last_num + 1
         num(i) := last_num
         j(i) := 1
      else if p(i) ≠ Γ_{j(i)}(i) then {back edge}
         add {i, Γ_{j(i)}(i)} to B
         j(i) := j(i) + 1
```

A depth-first search of an undirected graph $G = (V, E)$ partitions the edges in E into two sets T and B. An edge $\{v, w\}$ is placed in T if node w has not been previously visited when we are at node v considering edge $\{v, w\}$. Otherwise, edge $\{v, w\}$ is placed in B. The edges in T are called *tree edges*, and those in B *back edges*. The subgraph (V, T) is an undirected forest called *depth-first spanning forest* for G. Moreover, we number nodes in $|V|$ in the order they are visited during the depth-first search. The number of node v is denoted by $num(v)$. This procedure is summarized in Algorithm 6.1. If v was visited before w and if $\{v, w\} \in T$, v is called the *parent* of w and w is a *child* of v.

To check two-connectivity of a graph, we define for each node $v \in V$

$$\begin{aligned} inf(v) \ := \min \ & (\{num(v)\} \cup \{num(w) : \{v,w\} \in B\} \\ & \cup \{inf(w) : w \text{ is a child of } v\}). \end{aligned}$$

If a node v is not the root of the depth-first search, then it can be shown that v is an articulation point if and only if it has a child w such that $inf(w) \geq num(v)$. If v is the root of the depth-first search, then v is an articulation point if we backtrack to v before having visited all nodes (see Aho et al. [3] for details). Moreover, the computation of $inf(v)$ for all nodes can be embedded in the depth-first search algorithm, leading to a computing time of $O(m)$ to check two-connectivity of a graph.

We now turn back to the problem of checking the two-connectivity of G_K. If we apply the two-connectivity algorithm to G, edges that do not belong to any feasible cycle must be ignored in the search, since they do not belong to G_K. Moreover, if a back edge $\{v, w\} \in B$, examined from node v, is such that $num(w) \geq inf(v)$, then this edge is not necessary to ensure two-connectivity. Therefore, we do not need to check if it belongs to G_K, reducing the number of necessary checks.

To check if $e \in G_K$, we compute the shortest cycle using e in G. If the length of this cycle is less than or equal to K, all the edges belonging to the cycle are in G_K, and we can keep this information in order to avoid redundant checking. This is done by giving a mark *ok* to each edge which is set to true as soon as we have found a feasible cycle containing the edge. Moreover, the cycle may use a node v such that $num(v) < inf(w)$ for some w in the cycle, allowing us to decrease $inf(w)$ faster than allowed by the classical depth-first search approach, thus again reducing the amount of work needed. The subroutine used to check if an edge e belongs to G_K, together with the update of two-connectivity information and the *ok* marks, is described in Algorithm 6.2.

This subroutine must be called in the two-connectivity check for G_K if the edge $\{v, w\}$, which is examined from node v, has an *ok* mark not set to true, and the edge is a tree edge or a back edge such that $num(w) < inf(v)$. The complete procedure is described in Algorithm 6.3.

Algorithm 6.2 The function $check\_cycle(G, e, ok, inf, num)$

{This function is used in Algorithm 6.3 to check if an edge e belongs to G_K.}
compute the shortest cycle $C := (V_C, E_C)$ using e
if length(C) $\leq K$ **then**
  $minnum := \min_{k \in V_C} num(k)$
  for all $k \in V_C$ **do**
    $inf(k) := \min(inf(k), minnum)$
  for all $e \in E_C$ **do**
    $ok(e) :=$ true

Algorithm 6.2 requires the computation of a shortest path between the endpoints of an edge, and has thus a complexity of $O(n^2)$ if we use Dijkstra's algorithm. This subroutine is called at most m times, leading to a complexity of $O(mn^2)$ for Algorithm 6.3. Even if this worst-case complexity is the same as the simple approach described at the beginning of this section, it is more efficient in practice since the number of calls to Algorithm 6.2 is usually far less than m, while it would always be equal to m with the straightforward approach.

6.2 SEPARATION OF CUT CONSTRAINTS

The separation of cut constraints $x(\delta(W)) \geq 2$ can be carried out by computing a minimum cut in the graph, with capacities given by the current LP solution. This can be done in polynomial time, e.g. by the Gomory-Hu algorithm [46] that requires $n - 1$ maximum flow computations.

6.3 SEPARATION OF METRIC INEQUALITIES

The metric inequality corresponding to a set of node potentials $(\alpha_k)_{k \in V}$ satisfying $\alpha_i - \alpha_j > K - d(i,j)$ is

$$\sum_{f \in E \setminus \{e\}} v_f x_f \geq x_e,$$

Algorithm 6.3 Checking two-connectivity of G_K

Data : a graph $G = (V, E)$, a real number $K > 0$, a node $v \in V$.
Return value : *true* if G_K is two-connected, *false* otherwise.
Notation : $d(i)$: degree of i in G, $\Gamma_j(i)$: j-th node adjacent to i in G.

```
for all e ∈ E do
   ok(e) := false {ok(e) = true if e ∈ G_K}
for all i ∈ V do
   p(i) := −1; num(i) := ∞; inf(i) := ∞
p(v) := v; num(v) := 1; i := v; j(i) := 1; last_num := 1
while (j(i) ≤ d(i)) or (i ≠ v) do
   if j(i) = d(i) then {all the edges leaving i have been examined}
      Q := inf(i); i := p(i) {backtracking}
      if Q ≥ num(i) then
         if (i ≠ v) or (last_num < |V|) then {i art. point}
            return false
   else {we examine the next edge leaving i}
      inf(i) := min(inf(i), num(i))
      if p(Γ_j(i)(i)) = −1 then {tree edge}
         if ok({i, Γ_j(i)(i)}) = false then
            check_cycle(G, {i, Γ_j(i)(i)}, ok, inf)
         if ok({i, Γ_j(i)(i)}) = true then
            p(Γ_j(i)(i)) := i
            i := Γ_j(i)(i)
            last_num := last_num + 1
            num(i) := last_num
            j(i) := 1
      else if p(i) ≠ Γ_j(i)(i) then {back edge}
         if num(Γ_j(i)(i)) < inf(i) then
            if ok({i, Γ_j(i)(i)}) = true then
               inf(i) := num(Γ_j(i)(i))
            else
               check_cycle(G, {i, Γ_j(i)(i)}, ok, inf)
      j(i) := j(i) + 1
return true
```

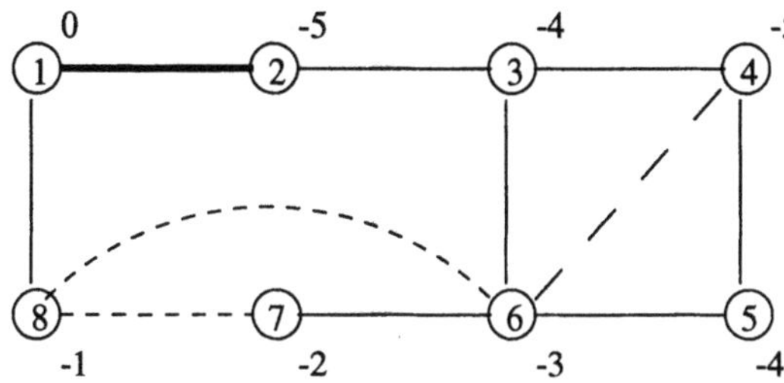

Figure 6.1. Values of $\alpha_k = \pi_k$

where

$$v_f = \min\left(1, \max\left(0, \frac{|\alpha_l - \alpha_k| - d(k,l)}{\alpha_i - \alpha_j + d(i,j) - K}\right)\right)$$

for all $f := \{k, l\} \in E \backslash \{e\}$.

Given a fractional point $(x_e)_{e \in E}$, an easy way to solve the separation problem for these inequalities is to compute the minimum cost flow defined by (4.10)-(4.13) for each $e \in E$ and to consider the inequality derived from α_k values equal to the corresponding node potentials π_k. However, it is possible to generate stronger inequalities as illustrated below.

Consider the graph depicted in Figure 6.1, with edge lengths equal to 1 and a bound $K = 5$. Edge $\{1,2\}$ violates the ring constraints. Let π_k be the node potentials corresponding to the minimum cost flow between 1 and 2. These values are given in Figure 6.1. The inequality obtained for edge $\{1, 2\}$ by taking $\alpha_k = \pi_k$ is

$$x_{68} + x_{46} \geq x_{12}. \tag{6.1}$$

The aim of introducing this inequality is to impose that the cost of the flow between nodes 1 and 2 becomes less than or equal to 4. If we set $x_{\{4,6\}} = 1$ in the solution, inequality (6.1) is satisfied while the cost of the flow remains equal to 4.5. Moreover, if we take $\alpha_4 = -4$ in place of $\alpha_4 = -5$, we get the new valid inequality

$$x_{68} \geq x_{12}, \tag{6.2}$$

which is stronger than (6.1) and which involves only edges that contribute to decrease the cost of the flow between nodes 1 and 2.

More precisely, we would like to find an inequality in which each involved edge contributes to the decrease of the cost of the flow between i and j if added to the solution. Algorithm 6.4 provides a heuristic to compute $(\alpha_k)_{k\in V}$ with this property. The aim of this heuristic is to reduce the node potentials given by the minimum cost flow by the mean of a backward Dijkstra-like procedure.

Note that the computation can be restricted to the subgraph Ω_e induced by edges belonging to a feasible cycle including e, i.e. $\Omega_e = \bigcup_{C\in\mathcal{C}_e} C\backslash\{e\}$. Applied to the example of Figure 6.1, this algorithm leads to inequality (6.2).

Algorithm 6.4 is a polynomial time procedure to separate metric inequalities. The inequalities generated provide a weaker linear relaxation than the one obtained using ring-cover inequalities, as shown in Proposition 4.20. However, if the algorithm stops with node potentials $(\pi_k)_{k\in V}$ such that $\pi_j + K - d_e \geq 0$ (line 4), then all ring-cover inequalities for edge e are satisfied (see Proposition 6.1), and the two linear relaxations yield the same optimal solution.

Applying this algorithm instead of an exact separation for ring-cover inequalities may fail if the algorithm stops with $\pi_j + K - d_e < 0$ and an inequality which is not violated (line 22). However, results obtained in practice show that this heuristic is very effective, and the linear relaxation values are close to each other.

PROPOSITION 6.1 *If Algorithm 6.4 applied to edge $e := \{i, j\} \in E$ finishes with $\pi_j + K - d_e \geq 0$, then all ring-cover inequalities (4.7) are satisfied.*

PROOF.

Since $-\pi_j \leq K - d_e$, the flow generated by the minimum cost flow algorithm uses only paths of length less than or equal to $K - d_e$. Each one of these paths, combined with edge e, forms a feasible cycle belonging to $\mathcal{C}_e$.

Algorithm 6.4 Separation of metric inequalities

Data : a graph $G = (V, E)$, a real number $K > 0$, a vector $\tilde{\mathbf{x}} := (\tilde{x}_f)_{f \in E}$, an edge $e := \{i, j\} \in E$.

Compute a minimum cost flow of value $\tilde{x}_e$ between i and j with capacities $(\tilde{x}_f)_{f \in E \setminus \{e\}}$ in $G - e$.
Let $(\pi_k)_{k \in V}$ be the corresponding node potentials.

if $\pi_j + K - d_e \geq 0$ **then**
    stop{all metric inequalities having x_e as right-hand side are satisfied}

for all $k \in V$ **do**
    if $k = j$ **then**
        $\alpha_k := \pi_j$
    else
        $\alpha_k := 0$
    label(k) := false

while $\exists k \in V$: label(k) = false **do**
    $l := \arg\min\{\alpha_k : \; k \in V,$ label(k)=false$\}$
    for all $k \in V$: $\{k, l\} \in E \setminus \{e\}$ **do**
        if $(\alpha_k > \alpha_l + d(k, l))$ and $(\alpha_l + d(k, l) \geq \pi_k)$ **then**
            $\alpha_k := \alpha_l + d(k, l)$
    label(l) := true

for all $f := \{k, l\} \in E \setminus \{e\}$ **do**
    $v_f := \min\left(1, \max\left(0, \frac{|\alpha_l - \alpha_k| - d(k,l)}{\alpha_i - \alpha_j + d(i,j) - K}\right)\right)$

if $\sum_{f \in E \setminus \{e\}} v_f \tilde{x}_f < \tilde{x}_e$ **then**
    $\{(\alpha_k)_{k \in V}$ define a violated metric inequality$\}$
else
    {no violated inequality was found}

By setting y_e^c equal to the flow sent on the corresponding path leads to a feasible solution of the system (4.3)-(4.4) such that $\sum_{c \in C_e} y_e^c \geq \tilde{x}_e$. We can conclude that all ring-cover inequalities are satisfied. □

6.4 SEPARATION OF SUBSET INEQUALITIES

The subset inequality induced by a subset of edges $S \subseteq E$ such that $(G - S)_K$ is not two-connected is $x(S) \geq 1$.

The separation of subset inequalities for the two-connected network problem ($K = \infty$) can be performed in polynomial time as facet-defining subset inequalities are node-cut constraints (see Proposition 4.14).

However, the general case is harder to solve. We want to find a subset $S \subseteq E$ such that $x(S) < 1$ and $(G - S)_K$ is not two-connected. Such a subset S does not necessarily define a node-cut and has no known special structure. The NP-completeness of this problem remains an open question, and we rely on the greedy heuristic described in Algorithm 6.5 to solve the separation problem.

The algorithm is divided in two phases : the first phase (lines 1-10) tries to find in a greedy way a subset $F \subseteq E$ such that $\tilde{x}(E \backslash F) < 1$ and F_K is not two-connected. If it succeeds, $x(E \backslash F) \geq 1$ is a violated subset inequality. The second phase (lines 11-18) is used to find $S \subseteq E \backslash F$ such that for all $e \in S$, $(G - (S \backslash \{e\}))_K$ is two-connected. The generated subset inequality is thus facet-defining by Theorem 4.13.

The main drawback of this heuristic is that it often fails to find a suitable F. It is easy to adapt the algorithm to perform the exact separation by a backtracking procedure, enumerating all the subsets F such that $\tilde{x}(E \backslash F) < 1$. However, there is an exponential number of such subsets, and complete enumeration is not efficient in practice. An intermediate approach is to allow a fixed number of backtracking steps. Our computational experiments show that performing 10 backtracking steps is a good tradeoff between computing time and quality of the separation.

Algorithm 6.5 Separation of subset inequalities

Data : a graph $G = (V, E)$, a real number $K > 0$, a vector $\tilde{\mathbf{x}} := (\tilde{x}_f)_{f \in E}$.

let $F := \{e \in E : \tilde{x}_f > 0\}$
if F_K is two-connected **then**
  sort edges of F in increasing order of $\tilde{x}_f$,
  such that $F = \{j(1), \ldots, j(|F|)\}$, $\tilde{x}_{j(i)} \le \tilde{x}_{j(k)}$ for all $i < k$.
  $i := 1$
  $v := 0$
  while $v + \tilde{x}_{j(i)} < 1$ and $i \le |F|$ and F_K is two-connected **do**
    $F := F \backslash \{j(i)\}$
    if F_K is two-connected **then**
      $v := v + \tilde{x}_{j(i)}$
    i := i + 1
if F_K is not two-connected **then**
  $S := \phi$
  for all $e \in E \backslash F$ **do**
    if $(F \cup \{e\})_K$ is two-connected **then**
      $S := S \cup \{e\}$
    else
      $F := F \cup \{e\}$
    $\{x(S) \ge 1$ is a violated facet-defining subset inequality$\}$
else
  {no violated subset was inequality found}

Another way of deriving violated subset inequalities comes from the separation of metric inequalities. Let $\sum_{f \in E \backslash \{e\}} v_f x_f \ge x_e$ be a metric inequality, and let $S := \{f \in E \backslash \{e\} : v_f > 0\}$. If $(G - S)_K$ is not two-connected and $x(S) < 1$, then S defines a violated subset inequality. This situation is often met in practice. Moreover, if $v_f = 1$ for all $f \in S$, then the subset inequality is stronger than the metric inequality.

6.5 SEPARATION OF RING-CUT INEQUALITIES

A ring-cut inequality is defined by a subset of nodes $W \subseteq V$, $\phi \neq W \neq V$ and an independent subset $S \subseteq \delta(W)$ in the ring-cut graph $RCG_{W,K}$. The inequality is given by

$$x(S) + 2x(\delta(W) \backslash S) \geq 3.$$

The separation of ring-cut inequalities is performed heuristically. Before describing the heuristic, we show that this separation problem is NP-complete.

PROBLEM 6.2 (RING-CUT-SEP) *Let $G = (V, E)$ be a graph, $K > 0$ a given constant and $\mathbf{x} := (x_e)_{e \in E}$ a vector such that $0 \leq x_e \leq 1$ for all $e \in E$.*

Does there exist a subset of nodes $W \subseteq V$, $\phi \neq W \neq V$, and an independent subset $S \subseteq \delta(W)$ in the ring-cut graph $RCG_{W,K}$, such that

$$x(S) + 2x(\delta(W) \backslash S) < 3 \ ?$$

THEOREM 6.3 *RING-CUT-SEP is NP-complete for any fixed $K \geq 3$.*

PROOF.

It is easy to see that RING-CUT-SEP belongs to NP. We show that the independent subset problem reduces to RING-CUT-SEP.

The independent subset problem is the following : given a graph $G = (V, E)$ and an integer p, $1 \leq p \leq |V|$, does there exists an independent subset $S \subseteq V$ of cardinality $|S| \geq p$?

The independent subset problem is known to be NP-complete (see Garey and Johnson [39]). Given an instance of this problem, we construct an instance of RING-CUT-SEP in the following way. Let $G' = (V_1 \cup V_2, E_1 \cup E_2 \cup \overline{E})$ be an undirected graph where V_1 and V_2 are two identical copies of V, E_1 and E_2 are edge sets defining complete graphs over V_1 and V_2, and $\overline{E}$ contains edges joining nodes in V_1 and V_2 corresponding to a same node in V. G' has $2|V|$ nodes and $|V|^2$ edges. This transformation is illustrated in Figure 6.2.

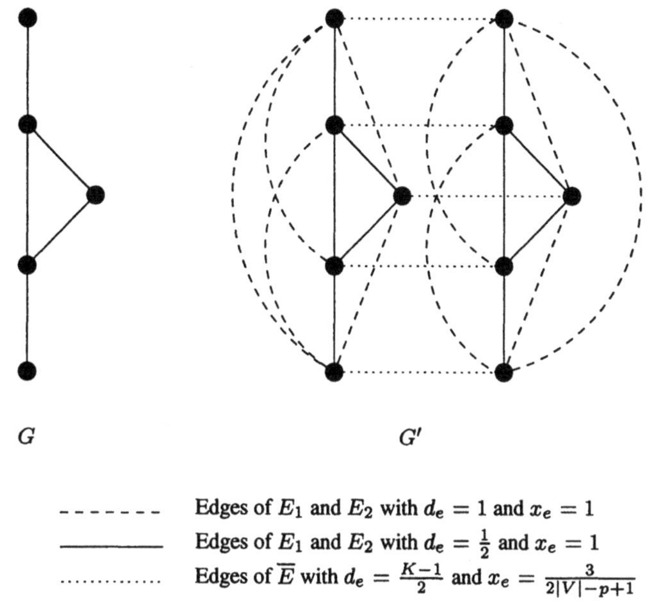

Figure 6.2. Transformation of independent subset into RING-CUT-SEP

The length of the edges and the vector $\mathbf{x}$ are defined as follows :

$$d_e = \begin{cases} 1 & \text{if } e \in E_1 \cup E_2 \\ & \text{and } e \text{ does not correspond to an edge in } E, \\ \frac{1}{2} & \text{if } e \in E_1 \cup E_2 \\ & \text{and } e \text{ corresponds to an edge in } E, \\ \frac{K-1}{2} & \text{if } e \in \overline{E}. \end{cases}$$

$$x_e = \begin{cases} 1 & \text{if } e \in E_1 \cup E_2, \\ \frac{3}{2|V|-p+1} & \text{if } e \in \overline{E}. \end{cases}$$

We must show that there exists an independent subset $S \subseteq V$ of cardinality $|S| \geq p$ in G if and only if there exists a subset of nodes $W \subseteq V_1 \cup V_2$, $\phi \neq W \neq V_1 \cup V_2$, and an independent subset $\overline{S} \subseteq \delta_{G'}(W)$ in the ring-cut graph $RCG_{W,K}$ of G', such that $x(\overline{S}) + 2x(\delta_{G'}(W) \backslash \overline{S}) < 3$.

Consider the cut defined by $W = V_1$. Each edge in this cut corresponds to a node in V. Let $e, f \in \delta_{G'}(W)$ be two edges in the cut. If there exists a feasible cycle containing e and f, it is easy to see that $(C_{e,f})_K = E_1 \cup E_2 \cup \{e, f\}$ and that $(C_{e,f})_K$ is two-connected, since E_1 and E_2 define complete subgraphs, since each edge in E_1 and E_2 has a length less than or equal to one, and since $K \geq 3$. A feasible cycle containing e and f must use at least one edge in E_1 and one edge in E_2. The only possibility is thus to use one edge of length $\frac{1}{2}$ in E_1 and in E_2, leading to a cycle of length equal to K. These edges may be chosen provided there exists an edge between the nodes corresponding to e and f in E. We can conclude that $(C_{e,f})_K$ is two-connected if and only if there is an edge in E between the nodes corresponding to e and f. Therefore, the ring-cut graph $RCG_{W,K}$ is isomorphic to G.

Suppose $S \subseteq V$ is an independent subset of cardinality $|S| \geq p$ in G, and let again $W = V_1$. Since $RCG_{W,K}$ is isomorphic to G, there exists $\overline{S} \subseteq \delta_{G'}(W)$ such that $|\overline{S}| = |S| \geq p$ and $\overline{S}$ is an independent subset in the ring-cut graph $RCG_{W,K}$. By the definition of $\mathbf{x}$, we obtain that

$$\begin{aligned} x(\overline{S}) + 2x(\delta_{G'}(W)\backslash\overline{S}) &= 2x(\delta_{G'}(W)) - x(\overline{S}) \\ &= \frac{3(2|V| - |\overline{S}|)}{2|V| - p + 1} \\ &\leq \frac{3(2|V| - p)}{2|V| - p + 1} \\ &< 3, \end{aligned}$$

and W and $\overline{S}$ define the requested violated ring-cut inequality.

If there exists a subset of nodes $W \subseteq V_1 \cup V_2$, $\phi \neq W \neq V_1 \cup V_2$, and an independent subset $\overline{S} \subseteq \delta_{G'}(W)$ in the ring-cut graph $RCG_{W,K}$ of G', such that $x(\overline{S}) + 2x(\delta_{G'}(W)\backslash\overline{S}) < 3$, then we must have $x(\delta_{G'}(W)) < 3$, since $\overline{S} \subseteq \delta_{G'}(W)$. The edge sets E_1 and E_2 define complete subgraphs with edges of value 1, and therefore each cut separating two nodes of V_1 or two nodes of V_2 has a value greater than or equal to $|V|$. The only possibility for W is then to have $\delta_{G'}(W) = \delta_{G'}(V_1)$, and the ring-cut graph $RCG_{W,K}$ is again isomorphic to G. Thus there exists an independent subset $S \subseteq V$ in G such that $|S| = |\overline{S}|$. Moreover,

$$3 > x(\overline{S}) + 2x(\delta_{G'}(W)\backslash\overline{S}) = \frac{3(2|V| - |\overline{S}|)}{2|V| - p + 1}.$$

It follows that $|S| = |\overline{S}| > p - 1$, thus $|S| \geq p$ and S is the requested independent subset. □

Since the separation problem for ring-cut inequalities is NP-hard, we solve it using a greedy heuristic. As we use the Gomory-Hu algorithm to separate cut constraints, we benefit from this information in the separation of ring-cut inequalities , considering the $n - 1$ minimum cuts provided by the Gomory-Hu tree as good candidates. To each cut of capacity less than 3 in the tree, we apply the greedy heuristic described in Algorithm 6.6 to determine an independent subset in the ring-cut graph corresponding to the cut.

Algorithm 6.6 Separation of ring-cut inequalities

Data : a graph $G = (V, E)$, a real number $K > 0$, a vector $\tilde{\mathbf{x}} := (\tilde{x}_f)_{f \in E}$, a subset $W \subseteq V$, $\phi \neq W \neq V$ such that $\tilde{x}(\delta(W)) < 3$.

$v := \tilde{x}(\delta(W))$
$S := \phi$
for all $e \in \delta(W)$ **do**
  $mark(e) :=$ false
while $(v < 3)$ and $(\exists e \in \delta(W) : mark(e) =$ false$)$ **do**
  $e := \arg\max\{\tilde{x}_f : f \in \delta(W), mark(f) =$ false $\}$
  for all $f \in \delta(W) : mark(f) =$ false **do**
    if $(C_{e,f})_K$ is two-connected **then**
      {edge $\{e, f\}$ belongs to the ring-cut graph}
      $v := v + \tilde{x}_f$
      $mark(f) :=$ true
  $mark(e) :=$ true
  $S := S \cup \{e\}$
if $v < 3$ **then**
  $\{x(S) + 2x(\delta(W) \backslash S) \geq 3$ is a violated ring-cut inequality$\}$

6.6 SEPARATION OF NODE-PARTITION INEQUALITIES

Given a node $z \in V$ and a partition $W_1, W_2, \ldots, W_p$ ($p \geq 2$) of $V \backslash \{z\}$, the corresponding node-partition inequality is

$$\frac{1}{2}\sum_{i=1}^{p} x(\delta_{G-z}(W_i)) \geq p - 1.$$

As described in Section 4.7, node-partition inequalities arise directly from the polytope of connected networks. This polytope is completely described by partition inequalities. The first separation algorithm for these inequalities was given by Cunningham [26] and requires $|E|$ min-cut computations. Barahona [6] reduced this computing time to $|V|$ min-cut computations.

Separating node-partition inequalities can thus be done for each node $z \in V$ by applying Barahona's algorithm to $G - z$. This requires $|V|^2$ min-cut computations.

In order to reduce this number, note that if G is a two-edge-connected network, any articulation point in G has a degree at least equal to 4. Therefore, we decided to apply Barahona's algorithm to $G - z$ only if $x(\delta(z)) > 3$, which leads to a much faster separation procedure.

6.7 SEPARATION OF WEIGHTED PARTITION INEQUALITIES

The separation of weighted partition inequalities can also be performed using Barahona's algorithm, as the weighted partition inequality (4.17) can be rewritten

$$\frac{1}{2}\sum_{i=1}^{p}\sum_{e \in \delta(W_i)} \frac{K - d_e}{K} x_e \geq p - 1.$$

Solving the separation problem for partition inequalities for the vector $\tilde{\mathbf{x}}$ defined by $\tilde{x}_e := \frac{K-d_e}{K} x_e$, $e \in E$ is thus equivalent to solving the separation problem for weighted partition inequalities for x.

6.8 SEPARATION OF CYCLOMATIC INEQUALITIES

Cyclomatic inequalities are again defined by a partition $W_1, W_2, \ldots, W_p$ $(p \geq 2)$ of V. The inequality is given by

$$\frac{1}{2}\sum_{i=1}^{p} x(\delta(W_i)) \geq M(p,K) = p + \min\left(\left\lceil \frac{p-K}{K-2} \right\rceil, \left\lceil \frac{p}{K-1} \right\rceil\right).$$

The main difficulty that appears when trying to separate cyclomatic inequalities is the fact that the right-hand-side of the inequality is not linear in the number p of subsets that define the partition.

To override this difficulty, we decided to approximate $M(p, K)$ by a linear function of the form $a(p-1)$. Therefore, we try to find a most violated inequality of the form

$$\frac{1}{2}\sum_{i=1}^{p} \sum_{e \in \delta(W_i)} \frac{x_e}{a} \geq p - 1,$$

which can again be performed using Barahona's algorithm for partition inequalities. If such a violated inequality is found, it is sufficient to check that the cyclomatic inequality defined by the same partition is also violated.

The effectiveness of this procedure depends heavily on the choice of a in order to have a good approximation of $M(p, K)$. Our choice was to take

$$a = \frac{nK}{(n-1)(K-1)}$$

which leads to $a(n-1) = n + \frac{n}{K-1}$ for $p = n$. Figures 6.3 and 6.4 illustrate this approximation for $K = 4$ and $K = 7$, respectively, and $n = 50$. We can observe that the approximation is quite accurate, and gives good results in practice.

6.9 IMPLEMENTATION OF THE BRANCH-AND-CUT ALGORITHM

In this section, we describe some strategic choices that were made in the implementation of our Branch-and-Cut algorithm for the 2CNBR problem. Our aim

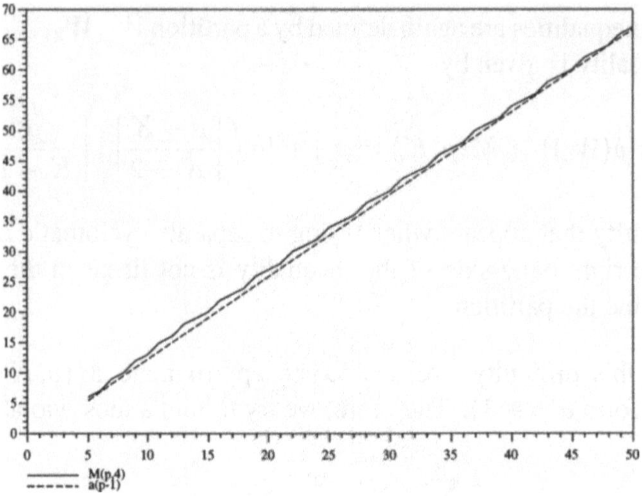

Figure 6.3. Approximation of $M(p, 4)$ for $n = 50$

here is not to describe in detail the general Branch-and-Cut framework, but to emphasize the problem-specific aspects of our algorithm. The outline of a generic Branch-and-Cut algorithm (for a minimization problem) is given in Algorithm 6.7. For a general introduction to Branch-and-Cut, we refer the reader to Thienel [78] or to Nemhauser and Wolsey [71] for a more complete survey on combinatorial optimization and polyhedral theory.

The algorithm was implemented in C++, using version 2.0 of the ABACUS library (Thienel [78, 79]), and CPLEX 4.0 as LP-solver. We present below the features of our implementation that differ from the default behavior of ABACUS, or that are problem-specific and need to be redefined for any application.

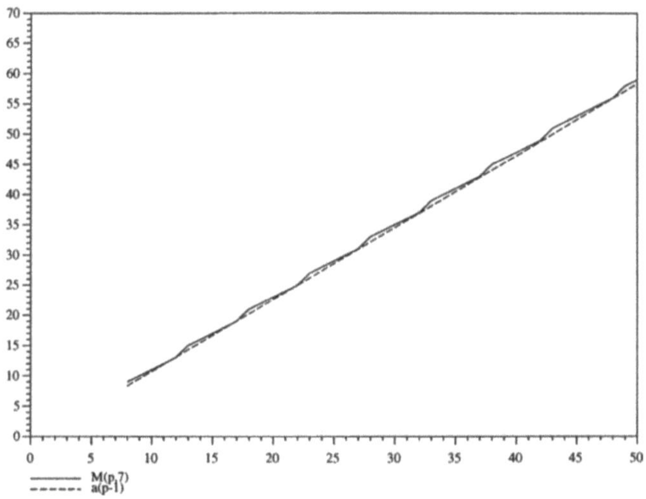

Figure 6.4. Approximation of $M(p, 7)$ for $n = 50$

Preprocessing

In our numerical experiments, we suppose for all instances that the original graph is complete, and therefore, the problem has $\frac{n(n-1)}{2}$ variables. However, it is possible that some of these edges have such a large length that they cannot belong to a feasible cycle. Such edges are removed in the preprocessing step, allowing a substantial decrease in the number of edges for instances where edge lengths are Euclidean.

Since we suppose edge lengths satisfy the triangle inequality, this preprocessing can be done in $O(n^3)$, checking for each triplet (i, j, k) of nodes if $d(i, j)+d(j, k)+d(k, i) \leq K$ and keeping only edges that belong to such a feasible triangle.

Unfortunately, this preprocessing is useless in the case of unit edge lengths, since any triangle is feasible when $K \geq 3$. As reported in Chapter 8, in this case, we fix a bound on the maximum cost of an edge in order to reduce the number of variables.

Algorithm 6.7 Branch-and-Cut

Notation : llb : local lower bound, glb : global lower bound, gub : global upper bound.

```
initialize first LP
solve LP; llb := value of the optimal solution
if llb ≥ gub then
    if all nodes have been processed then
        stop{optimization completed}
    else
        select an unprocessed node
        goto 2
else
    separate inequalities
    if violated inequalities are found then
        add inequalities to LP; goto 2
    else if fractional solution then
        branch on a variable with fractional value
        goto 2
    else
        set gub = llb
        if all nodes have been processed then
            stop{optimization completed}
        else
            select an unprocessed node
            goto 2
```

Initial LP

To start the optimization, we have to define the constraints of the first linear program. Our choice was to take only the degree constraints $x(\delta(v)) \geq 2$ for all $v \in V$, leading to the following initial LP :

$$\min \sum_{e \in E} c_e x_e$$

s.t.

$$x(\delta(v)) \geq 2 \qquad v \in V,$$
$$0 \leq x_e \leq 1 \qquad e \in E.$$

Moreover, in the case of unit edge lengths, we also introduce the lower bound on the number of edges $x(E) \geq M(n, K)$.

Upper bound computation

An important issue in the effectiveness of a Branch-and-Cut algorithm is the computation of good upper bounds. Due to the effectiveness of the Tabu Search heuristic presented in Section 7.5, we perform 600 iterations of Tabu Search in parallel with the Branch-and-Cut algorithm to obtain a good upper bound.

Moreover, we try to transform each LP-solution obtained in the Branch-and-Cut to a feasible solution by rounding up to 1 all the variables with fractional value.

Cutting plane generation

The separation of valid inequalities is performed as follows. We first separate inequalities from the standard pool (see below). If all the inequalities in the pool are satisfied by the current LP-solution, we separate the classes of inequalities introduced in the previous chapters in the following order :

1. weighted partition inequalities (or cyclomatic inequalities);
2. subset inequalities;
3. cut constraints;
4. ring-cut inequalities;
5. node-partition inequalities;
6. metric inequalities.

This order was chosen after a series of numerical experiments, the choice of inequalities separated first seemingly being the best trade-off between separation time and efficiency of the cuts.

Moreover, we go to the next class of inequalities only if the number of generated cutting planes is less than 50. Otherwise, we solve the LP again and restart the separation procedure.

All inequalities are global (i.e. valid in all the tree), except ring-cut inequalities that are valid locally. This is due to the fact that setting a variable to 0 may change the ring-cut graph generated by a cut not containing the corresponding edge.

Pool management

The pool used to store generated inequalities is the standard pool in ABACUS. We start with a pool size equal to 100 times the number of nodes in the network, and we allow this size to be increased dynamically if necessary. All the generated inequalities are put in the pool and are dynamic, i.e. they are removed from the current LP when they are not active.

Separation vs. branching

The separation of valid inequalities is time-consuming, and choosing a good tradeoff between separation and branching is often an important issue in the effectiveness of a Branch-and-Cut algorithm. Our strategy is the following : in the first nodes of the Branch-and-Bound tree, we perform separation until no more violated inequality is found, in order to obtain the best possible lower bound. In subsequent nodes, we limit the cutting plane generation to 5 iterations, since we observed that the cuts generated become less effective in improving the lower bound. We limit the number of cutting plane generations for all nodes that are at a level deeper than 7 in the Branch-and-Bound tree.

Enumeration strategy

The most often used enumeration strategy in a Branch-and-Cut algorithm is the *best-first search* strategy, in which the node with the lowest lower bound is optimized first. This has the advantage of improving the lower bound rapidly, but has the major drawback of being very memory-consuming, since the number of unexplored nodes in the tree is also growing rapidly. Our numerical experiments showed that the number of nodes in our Branch-and-Bound tree grows fast with the size of the problem, and the best-first search approach becomes impractical as soon as the size of the tree exceeds 20,000 nodes, due to the large memory necessary.

Besides, the *depth-first search* strategy optimizes first the deepest node in the tree. This has the advantage that the number of unexplored nodes is bounded by the number of variables, thus limiting the amount of consumed memory. Unfortunately, the lower bound increases much slower, and remains equal to the LP-relaxation (i.e. the bound obtained at the first node of the tree) for a long time. If we wish to run the Branch-and-Cut algorithm for a limited time, the lower bound obtained is really bad, and of no practical utility.

We therefore chose to implement the following mixed strategy : we apply best-first search, but we switch to depth-first search each time the number of unexplored nodes becomes greater than 1000. In this way, the lower bound increases rapidly at the beginning of the optimization, and the amount of memory consumed stays bounded.

6.10 SUMMARY OF POLYHEDRAL RESULTS

To conclude the polyhedral study of the 2CNBR problem, we summarize in Table 6.1, for each class of inequalities, facet properties and results concerning the separation problem. Numbers in parenthesis refer to the corresponding theorems.

Name	Form	Facet results	Separation
Cut	$x(\delta(W)) \geq 2$	NSC (4.10)	Exact, in polynomial time
Subset	$x(S) \geq 1$, where $(G-S)_K$ is not two-connected	NSC (4.13)	Heuristic, complexity unknown
Ring-cut	$x(S) + 2x(\delta(W)\backslash S) \geq 3$, where S is an independent subset in the ring-cut graph	NC (4.16) SC (4.17,4.18)	Heuristic, NP-complete
Ring-cover	$\sum_{f\in E\backslash\{e\}} \overline{u}_f x_f \geq x_e$, where $\overline{u}_f$ satisfy (4.5)-(4.6)	?	Not implemented, NP-complete
Metric	$\sum_{f\in E\backslash\{e\}} v_f x_f \geq x_e$ where v_f satisfy (4.9)	?	Heuristic, complexity unknown
Node-partition	$\frac{1}{2}\sum_{i=1}^{p} x(\delta_{G-z}(W_i)) \geq p-1$	SC (4.21)	Exact, in polynomial time
Weighted partition	$\frac{1}{2}\sum_{i=1}^{p} \sum_{e\in\delta(W_i)} (K-d_e)x_e \geq (p-1)K$	?	Exact, in polynomial time
Cyclomatic	$\frac{1}{2}\sum_{i=1}^{p} x(\delta(W_i)) \geq M(p,K)$	G complete, $p=n$ (5.6)	Heuristic, complexity unknown

NSC : Nessary and sufficient conditions
SC : Sufficient conditions
NC : Nessary conditions

Table 6.1. A summary of valid inequalities

Chapter 7

HEURISTICS

In this chapter, we describe five heuristics used to generate feasible solutions; the first four are constructive. The last one is a Tabu Search heuristic that was used in conjunction with the Branch-and-Cut algorithm.

In each case, we assume that the problem is feasible, i.e. the set of potential edges is two-connected and satisfies the ring constraints.

7.1 EAR-INSERTING METHOD

The basic idea of this heuristic is derived from the ear-inserting procedure proposed by Monma and Shallcross [69] for designing minimum-weight two-connected networks. We adapt this procedure in order to satisfy ring constraints. It is developed for the particular case where the edge costs are equal to their lengths.

Our aim is to construct cycles with length at most K, so that the total cost does not increase too much: we first construct the minimum length cycle; we then add at each step the node which minimizes the total length increase. When no new node can be added to the starting cycle without violating the ring constraints, we create a new cycle including a pair of nodes already in the solution – in order to preserve the two-connectivity requirement. We add as many nodes as possible to this cycle, until we reach the bound. Then we repeat the procedure by initiating a new cycle until all nodes are in the solution. Moreover, we also start a new cycle when it is better than augmenting the current one. This procedure is improved by local 2-opt exchange in the cycle being constructed and by allowing re-use of a node already in the solution if it leads to a better solution.

A detailed description of this heuristic is provided in Algorithm 7.1.

7.2 CUTTING CYCLES IN TWO EQUAL PARTS

Our numerical experiments show that the optimal two-connected network with no further constraints turns out to be often a Hamiltonian cycle. By adding edges to this cycle, we can satisfy the ring constraints. In this heuristic, edges are added to cut the cycle in two almost equal parts, and the cutting is repeated for all sub-cycles until the ring constraints are satisfied.

The starting Hamiltonian cycle is created by a greedy procedure followed by 2-opt and 3-opt exchanges. The choice of the chord to add at each step is driven by the quality of the division in two and the cost of the edge to add. A parameter controls this choice and is updated in order to give more importance to exact cutting as the cycle sizes decrease. A detailed description is reported in Algorithm 7.2.

Algorithm 7.1 Ear-inserting heuristic

Data : a graph $G = (V, E)$ such that $c(i,j) = d(i,j)$ for all $i, j \in V$, a real number $K > 0$.

1: Create a minimum length cycle.
2: Select $i, j, k \in V$ such that :
- i, j are adjacent and belong to the current cycle,
- k does not belong to the current solution,
- $c(i,k) + c(k,j) - c(i,j)$ is minimal.

3: If the length of the cycle induced by replacing edge $\{i,j\}$ by edges $\{i,k\}$ and $\{k,j\}$ is less than or equal to K, make the change and repeat the procedure until no more node can be added.
4: **repeat**
5: Select $i, j, k \in V$ such that :
- i, j are in the current solution,
- k does not belong to the current solution,
- $c(i,k)+c(k,j)+c(i,j)(1-x_{ij})$ is minimal, where x is the incidence vector of the current solution.

6: Add edges $\{i,k\}$, $\{k,j\}$ and, if necessary, $\{i,j\}$.
7: Select $l, m, k \in V$ such that :
- l, m are adjacent, belong to the current cycle and l, $m \notin \{i, j\}$, where i and j are the ear extremities chosen at line 5,
- k is of degree 0 or 2,
- $c(s,k) + c(k,t) - c(s,t) > c(l,k) + c(k,m) - c(l,m)$, if k is of degree 2, with neighbors s and t,
- $c(l,k) + c(k,m) - c(l,m)$ is minimal.

8: **until** all nodes are in the solution

7.3 PATH-FOLLOWING METHOD

In this method, we first determine a Hamiltonian tour. Then, beginning with any node, we follow the tour in, say, the clockwise direction, until the most distant node of the tour, such that connecting it with the starting node yields a cycle of length not greater than K. We insert that edge and start again taking as first node the one that was considered last in the previous iteration, until the solution graph becomes

Algorithm 7.2 Cut-in-two heuristic

Data : a graph $G = (V, E)$, a real number $K > 0$.

Create a starting Hamiltonian cycle
(e.g. greedy procedure + 2-opt + 3-opt).
Set $z := 1$.
repeat
  for all cycles of length $> K$ in the solution **do**
    Add in the current solution the chord $\{k, l\}$ in the current cycle that minimizes

$$z\, c(k,l) - (1 - z)\, \delta$$

    where δ = length of the current cycle - length of the shortest of the two cycles induced by the chord /2.
    $z := z/1.33$
until all cycles have a length $\leq K$

feasible. The procedure is repeated starting from every node and the best solution is kept.

Although rarely attractive in practice, this simple constructive heuristic has the interesting feature of suggesting a polynomial algorithm with a worst-case performance bound with respect to the cost of an optimal solution, in the important particular case that the costs satisfy the triangle inequality (and hence G is a complete graph).

Assume we apply the polynomial heuristic starting with the Hamiltonian tour obtained by Christofides heuristic [18]. Let $\hat{c}_H$ be the cost of this tour and $\hat{c}_{2CNBR}$ the cost of the solution to the 2CNBR problem found by the path-following heuristic. Because of the triangle inequality, we have $\hat{c}_{2CNBR} \leq 2\hat{c}_H$. It was already observed by Frederickson and Já Já [36] that $\hat{c}_H \leq \frac{3}{2} c^*_{TC}$ where c^*_{TC} is the cost of an optimal two-connected subgraph of G. Obviously, $c^*_{TC} \leq c^*_{2CNBR}$, the cost of an optimal solution to the 2CNBR problem. Hence, this version of the path-following heuristic gives a solution with a cost never worse than 3 times the optimal cost.

Note that each inequality used to get that bound is tight. However, examples for which the bounds are reached do not coincide, and the question of the tightness of our worst-case bound remains open.

7.4 STINGY METHOD

The fourth heuristic is based on a greedy removal of edges. It starts with all edges and tries to remove the most costly one while preserving feasibility. A detailed description is given in Algorithm 7.3.

Algorithm 7.3 Stingy heuristic

Data : a graph $G = (V, E)$, a real number $K > 0$.

```
sort edges of E in increasing order of cost c_f,
   such that E = {j(1),...,j(m)}, c_j(i) ≤ c_j(k) for all i < k.
let F := E
for k := m to 1 do
   if (F\{j(k)})_K is two-connected then
      F := F\{j(k)}
return F
```

7.5 TABU SEARCH

Among local search heuristics, Tabu Search has proved its effectiveness in many applications. It was introduced by Glover [40] and since then, it has been applied to hundreds of combinatorial optimization problems. Basic concepts of Tabu Search and extensive surveys on these applications can be found in Glover and Laguna [41] and Hertz et al. [54].

Suppose we want to minimize a function f over a set X of feasible solutions. Local search techniques are iterative procedures in which a neighborhood $\mathcal{N}(x)$ is defined for each feasible solution $x \in X$, and which search the next solution x' in the neighborhood $\mathcal{N}(x)$ of the current solution x. Among these methods, the simpler and most famous is the descent method which stops at a local minimum. This method is summarized in Algorithm 7.4.

Algorithm 7.4 Descent method

```
choose an initial solution x ∈ X.
find x' ∈ N(x) such that f(x') ≤ f(y) for all y ∈ N(x).
if f(x') ≥ f(x) then
    stop
else
    x := x'
    go to Step 2.
```

Note that the stingy heuristic presented in the previous section belongs to this class of methods : the initial solution is the complete graph, and the neighbors of a solution are defined by removing one single edge from this solution.

To override the fact that the descent method stops at a local minimum, several meta-heuristics have been proposed. Most of these differ in the way the neighborhood of the current solution is explored. The most famous of these meta-heuristics are simulated annealing, genetic algorithms and Tabu Search. For a complete survey about local search techniques, we refer the reader to Aarts and Lenstra [1].

To avoid termination at a local minimum, meta-heuristics allow nonimproving moves. However, this presents the risk of visiting a solution again and more generally of cycling. In Tabu Search, the use of a *short-term memory* helps to forbid moves that might lead to recently visited solutions. With such a memory, the structure of $\mathcal{N}(x)$ will depend upon the itinerary of the search, hence upon the iteration k. Therefore, we refer to $\mathcal{N}(x,k)$ instead of $\mathcal{N}(x)$. Some recently visited solutions (considered as *tabu solutions*), must be removed from $\mathcal{N}(x,k)$. The simplest way to implement this is to keep a list T (*tabu list*) of the last $|T|$ visited solutions, and to define $\mathcal{N}(x,k) := \mathcal{N}(x) - T$. However, such a list may be impractical to use (e.g. too memory-consuming). Therefore, one often considers the exploration process as a series of *moves* from one solution to the next. For each solution x, we define $M(x)$ as the set of moves m that can be applied to x in order to obtain a new solution x'. This will be denoted by $x' = x \oplus m$, and $\mathcal{N}(x) = \{x \oplus m : m \in M(x)\}$. In general, moves that are considered are reversible, i.e. for each $m \in M(x)$, there exists a move m^{-1} such that $(x \oplus m) \oplus m^{-1} = x$. Instead of keeping in T the last $|T|$ solutions, one may therefore only keep track of the last $|T|$ moves or the last $|T|$ reverse moves associated to the moves actually performed.

The main drawback of this simplification of the tabu list is that we may give a tabu status to a solution that may be unvisited so far. Some relaxation of the tabu status is then necessary : we overrule the tabu status when some tabu solution looks attractive. This attractiveness is measured by mean of an *aspiration criterion* that allows to consider a solution, even if it is tabu. For example, we would like to keep a tabu solution if its cost is better than the cost of the best solution found so far.

Another important issue for effectiveness is choosing an adequate length for the tabu list T. Instead of fixing this length, an alternative strategy is often used, which provides diversification as well as another mean of avoiding cycling. It consists in associating a randomly chosen *tabu tenure* to every move performed, the tabu tenure being to the number of iterations the reverse move has a tabu status.

Many other ingredients (often problem-specific) may appear in an effective tabu search algorithm. Nevertheless, basic tools that were presented here are sufficient in the scope of this work, and we refer the reader to the excellent surveys mentioned before for further details.

We are now ready to describe how we adapted the ideas of Tabu Search to the 2CNBR problem. Consider a minimal solution F (in the sense that removing any edge from F destroys feasibility). Such a solution can be obtained, for example, by the stingy heuristic described earlier. If a better solution F' exists, it is easy to see that F' contains at least one edge which is not in F. Based on this observation, we define the neighborhood of a solution F as the set of solutions F' that are obtained by applying to F the two following operations :

1. add an edge $e \in E \backslash F$ to F, leading to $F' := F \cup \{e\}$;
2. apply stingy to F', not allowing the removal of e.

By applying these operations, we obtain one new feasible solution for any $e \in E \backslash F$. To avoid cycling, we give a tabu status to the edge that was added to define the next iterate, and we first try to remove edges that are not tabu in the stingy phase. If an edge which is tabu was removed, the solution is considered as tabu and we accept it only if it improves the best cost found so far. This is the only aspiration criterion we use. The tabu tenure of each edge is randomly chosen between 4 and 10. Note that trying to remove edges which are not tabu first leads to a neighborhood structure strongly related to the current tabu set, therefore enforcing diversification. The complete heuristic is described in Algorithm 7.5.

Algorithm 7.5 Tabu Search heuristic

Data : a graph $G = (V, E)$, a real number $K > 0$, a feasible solution F_0.

```
sort edges of E in increasing order of cost c_f,
such that E = {j(1), ..., j(m)}, c_j(i) ≤ c_j(k) for all i < k.
i := 0
let F(i) := F_0; F* := F_0
for all e ∈ E do
   t(e) := 0 {tabu tenure}
while i < max_iterations do
   iter_cost := +∞
   for all e ∈ E\F(i) do {neighborhood exploration}
      F := F(i) ∪ {e}
      tabu := false
      for k := m to 1 do {stingy}
         if t(j(k)) ≤ i and j(k) ∈ F(i) then
            if (F\{j(k)})_K is two-connected then
               F := F\{j(k)}
      for k := m to 1 do {stingy on tabu edges}
         if t(j(k)) > i and j(k) ∈ F(i) then
            if (F\{j(k)})_K is two-connected then
               F := F\{j(k)}
               tabu := true
      if ((not tabu) and (c(F) < iter_cost )) or (c(F) < c(F*)) then
         F̄ := F
         f := e
         iter_cost := c(F)
   if iter_cost < +∞ then {the next solution was found}
      i := i + 1
      F(i) := F̄
      choose r at random between 4 and 10
      t(f) := i + r
      if c(F(i)) < c(F*) then
         F* := F(i)
   else {all solutions are tabu}
      for all e ∈ E do
         t(e) := 0 {reset tabu tenure}
return F*
```

Chapter 8

COMPUTATIONAL RESULTS

We present in this chapter numerical results obtained for the 2CNBR problem, with the Branch-and-Cut algorithm as well as the different heuristics presented in Chapter 7.

The Branch-and-Cut was implemented using ABACUS 2.0 and CPLEX 4.0, and tested on a SUN Sparc Ultra 1 workstation with a 166 Mhz processor and 128 Mo RAM. We fixed the maximum CPU time to 10 hours, except for randomly generated problems with 40 and 50 nodes, where it was limited to 3 hours, due to the large number of problems to solve. Moreover, for these large problems, we noticed that the bounds did not improve much after 3 hours. Heuristics were tested on a Pentium Pro 200 Mhz with 128 Mo RAM, running under Linux.

Since the number of instances solved is large (about 1000), we only present here a summary of results obtained. Detailed tables for all problems can be found in Appendix A.

Two classes of problems are solved, defined by different edge lengths. In the first class, edge lengths are equal to the rounded Euclidean distances. The second class is composed of problems with unit edge lengths. In all cases, we consider costs equal to the rounded Euclidean distance.

Tests were made for different values of the bound, for instances coming from real applications, with 12, 17, 30 and 52 nodes, and for random problems with nodes uniformly generated in a square of size 250×250. Random problems with 10 to 50 nodes were generated, and we tested five instances of each size.

Data on the randomly generated test problems are available at the Web page `http://smg.ulb.ac.be/~bfortz/2cnbm/data.html`.

When edge lengths are equal to the rounded Euclidean distances, we apply the preprocessing presented in Section 6.9, allowing to reduce the number of edges.

In the case of unit lengths, we already mentioned that the preprocessing does not allow to remove edges, and we are therefore left with a complete graph. To overcome this drawback, it seems quite natural to avoid edges with a too large cost. We therefore decided to suppress from the graph all edges having a cost greater than 80. If the resulting graph is infeasible, we increase this maximal cost by 20 until the graph becomes feasible.

8.1 BRANCH-AND-CUT FOR EUCLIDEAN EDGE LENGTHS

This section is devoted to the results obtained by the Branch-and-Cut algorithm for instances in which the edge lengths are equal to the rounded Euclidean distances.

Table 8.2 reports results obtained for problems coming from real applications, while Table 8.4 reports means of results obtained for randomly generated problems. The gaps are relative to the best upper bound found. The CPU times reported do not include the Tabu Search procedure. Only feasible problems appear in the tables. We also report in Table 8.3 the number of inequalities generated for each class of inequalities implemented, for the real instance with 52 nodes. Figure 8.1 and Figure 8.2 present two optimal solutions found on the real instances with 30 nodes.

For 20 nodes or less, all problems could be solved to optimality. For larger problems, we remark that the problems with a small value of K are much harder, the lower bound at the root of the Branch-and-Bound tree being far from the optimum.

Problems with a large value of K are easier to solve due to the fact that these problems are closer to the two-connected network problem (without ring constraints), which can be solved efficiently using cut and node-partition inequalities (for the instances we considered).

Abbreviations used in the tables are summarized in Table 8.1.

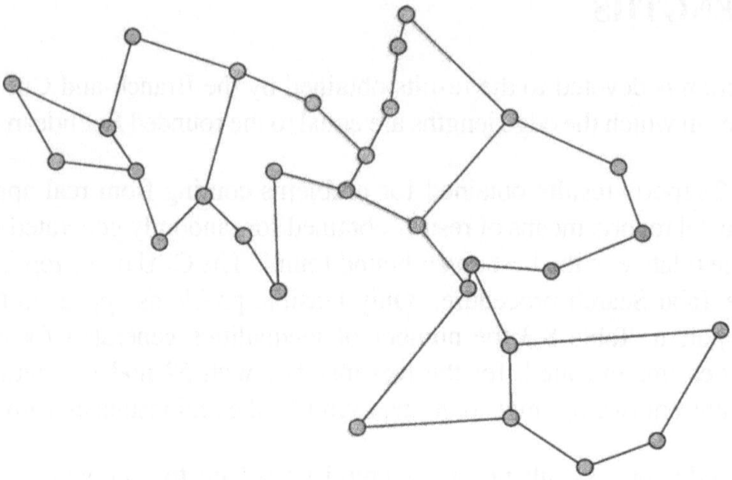

Figure 8.1. Optimal solution, 30 nodes real instance, $K = 200$

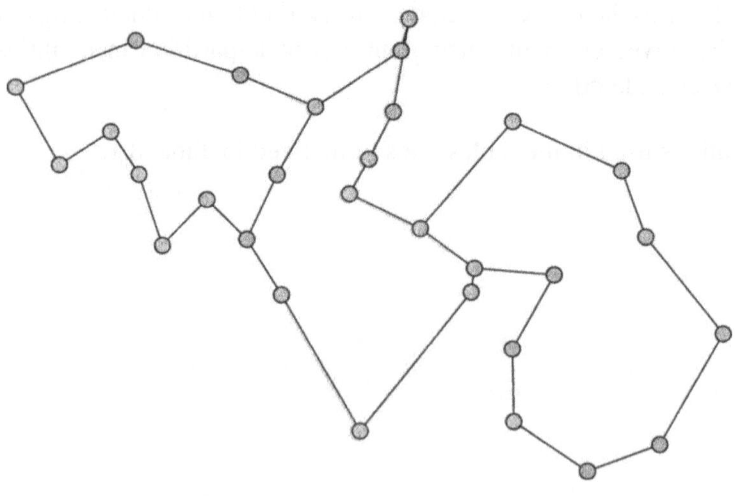

Figure 8.2. Optimal solution, 30 nodes real instance, $K = 350$

$\|V\|$	number of nodes in the graph
$\|E\|$	number of edges after preprocessing
K	bound on the length of cycles
p/o	for random problems, total number of problems/number of problems solved to optimality
# ineq.	number of inequalities generated
# B&B nodes	number of Branch-and-Bound nodes examined
LB (root)	lower bound obtained at the root node of the Branch-and-Bound tree
LB (final)	global lower bound at the end of the optimization
UB	best upper bound found
Gap	gap between the final upper and lower bounds : $gap = \frac{100(UB-LB)}{LB}$
CPU time	time spent in the Branch-and-Cut (without the Tabu Search)

Table 8.1. List of abbreviations

$\lvert V \rvert$	K	$\lvert E \rvert$	# ineq.	# B&B nodes	LB (root)	LB (final)	UB	Gap (%)	CPU time (hh:mm:ss)
12	200	49	220	35	577	622	622	0.0	0:00:09
12	250	58	64	7	530	541	541	0.0	0:00:01
12	300	63	125	35	514	541	541	0.0	0:00:27
12	350	65	65	5	509	521	521	0.0	0:00:01
12	400	66	92	13	503	521	521	0.0	0:00:02
12	450	66	79	15	499	516	516	0.0	0:00:12
12	500	66	27	1	496	496	496	0.0	0:00:01
17	200	88	296	39	802	834	834	0.0	0:00:22
17	250	109	379	103	743	789	789	0.0	0:00:47
17	300	119	144	23	711	726	726	0.0	0:00:08
17	350	126	223	45	701	725	725	0.0	0:00:14
17	400	133	336	139	688	720	720	0.0	0:00:36
17	450	135	69	7	680	689	689	0.0	0:00:03
17	500	136	93	11	675	689	689	0.0	0:00:05
30	150	174	34851	44411	1057	1180	1258	6.6	10:00:00
30	200	266	25477	32175	973	1065	1065	0.0	8:26:55
30	250	328	587	155	898	937	937	0.0	0:06:27
30	300	372	1090	815	875	898	898	0.0	0:15:53
30	350	406	390	115	858	874	874	0.0	0:03:51
30	400	422	559	365	849	868	868	0.0	0:07:04
30	450	432	1622	1313	843	868	868	0.0	0:26:19
30	500	435	273	47	837	844	844	0.0	0:01:45
52	150	559	7877	12127	1198	1249	1339	7.2	10:00:00
52	200	821	6514	7475	1132	1165	1239	6.4	10:00:00
52	250	1040	5161	4845	1097	1125	1184	5.2	10:00:00
52	300	1149	5220	3627	1073	1096	1138	3.8	10:00:00
52	350	1230	4486	3727	1061	1082	1085	0.3	10:00:00
52	400	1287	2828	1407	1048	1066	1066	0.0	4:32:55
52	450	1313	926	195	1042	1053	1053	0.0	0:54:33
52	500	1324	1558	417	1039	1053	1053	0.0	1:49:12

Table 8.2. Branch-and-Cut results, real applications, Euclidean edge lengths

K	cut	subset	ring-cut	node-partition	weighted partition	metric
150	143	250	2317	550	1151	3466
200	194	473	1074	364	481	3928
250	140	415	648	119	242	3597
300	239	561	383	117	286	3634
350	198	559	271	89	209	3160
400	170	466	115	41	101	1935
450	111	138	27	2	40	609
500	110	349	41	9	58	991

Table 8.3. Inequalities generated for the 52 nodes real instance, Euclidean edge lengths

$\|V\|$	$\|E\|$	K	p/o	# ineq.	# B&B nodes	Gap (root)	Gap (final)	CPU time (hh:mm:ss)
10	23.0	300	2/2	94.0	10.0	6.9	0.0	0:00:02
10	28.8	350	4/4	125.0	15.0	7.7	0.0	0:00:02
10	34.6	400	5/5	197.0	28.2	10.3	0.0	0:00:05
10	40.0	450	5/5	291.2	48.6	10.2	0.0	0:00:09
10	43.2	500	5/5	231.2	45.0	10.1	0.0	0:00:07
20	65.8	200	4/4	441.0	129.0	8.1	0.0	0:01:25
20	96.5	250	4/4	1942.2	2163.0	14.0	0.0	0:14:58
20	121.8	300	5/5	10059.2	14358.2	9.7	0.0	1:31:59
20	149.6	350	5/5	1307.6	1496.2	9.3	0.0	0:11:02
20	170.2	400	5/5	1123.6	685.8	9.0	0.0	0:07:28
20	182.4	450	5/5	2868.6	2734.6	8.0	0.0	0:26:06
20	187.2	500	5/5	679.0	208.2	6.1	0.0	0:03:03
30	145.7	200	3/2	17464.7	27474.3	14.3	1.9	6:41:46
30	226.2	250	4/0	26095.0	31736.0	18.6	9.2	10:00:00
30	288.8	300	5/2	17840.8	19965.8	12.6	4.9	7:12:18
30	345.8	350	5/4	7251.2	8892.6	7.4	1.0	3:19:02
30	381.6	400	5/4	4984.0	5570.6	5.3	0.5	2:25:00
30	414.8	450	5/5	1843.2	1605.4	5.1	0.0	0:48:41
30	426.8	500	5/5	1683.0	1171.0	4.9	0.0	0:39:38
40	258.0	200	4/0	5086.0	5454.0	17.7	9.6	3:00:00
40	368.8	250	5/0	4586.8	4677.0	18.0	10.6	3:00:00
40	466.6	300	5/0	3465.0	3319.0	15.1	9.1	3:00:00
40	561.6	350	5/0	3107.4	2854.2	10.3	5.3	3:00:00
40	645.6	400	5/0	3272.0	2873.4	6.9	2.6	3:00:00
40	719.8	450	5/1	2479.4	1640.2	7.0	3.1	2:45:33
40	753.6	500	5/0	2538.2	1610.2	5.8	1.9	3:00:00
50	243.0	150	1/0	3566.0	4917.0	9.4	3.9	3:00:00
50	442.2	200	4/0	2899.8	2368.0	16.4	10.8	3:00:00
50	607.0	250	5/0	1977.0	1531.4	15.7	11.0	3:00:00
50	775.4	300	5/0	1810.6	1011.8	9.4	6.0	3:00:00
50	926.0	350	5/0	1668.0	537.8	8.3	5.3	3:00:00
50	1055.2	400	5/0	1772.8	823.0	7.3	4.7	3:00:00
50	1141.0	450	5/0	1613.2	548.6	7.6	5.1	3:00:00
50	1195.4	500	5/0	1545.8	344.2	6.5	4.4	3:00:00

Table 8.4. Branch-and-Cut results, random networks, Euclidean edge lengths

8.2 BRANCH-AND-CUT FOR UNIT EDGE LENGTHS

Conclusions that can be derived from the results obtained by the Branch-and-Cut for instances in which edge lengths are unitary are essentially the same as in the previous section. The most difficult cases seem to be for K between 3 and 5, and especially for $K = 4$.

Table 8.5 reports results obtained for problems coming from real applications, while Table 8.7 reports the means of results obtained for randomly generated problems. We also report in Table 8.6 the number of inequalities generated for each class of inequalities implemented, for the real instance with 52 nodes. Figure 8.3 and Figure 8.4 present two optimal solutions found on the real instances with 30 nodes.

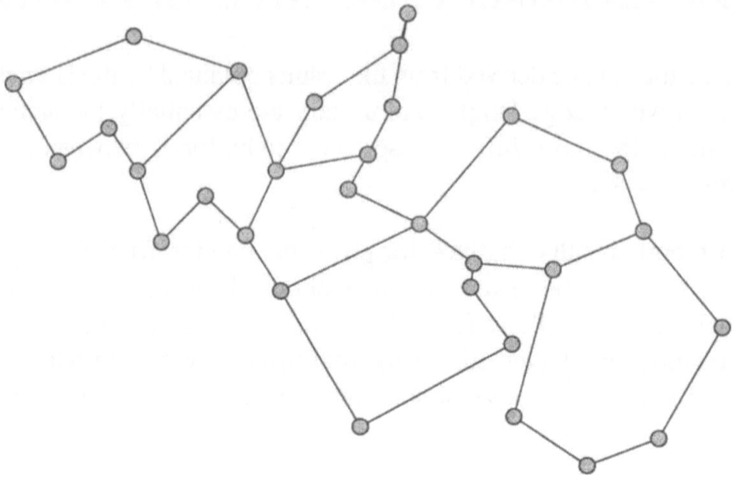

Figure 8.3. Optimal solution, 30 nodes real instance, $K = 6$

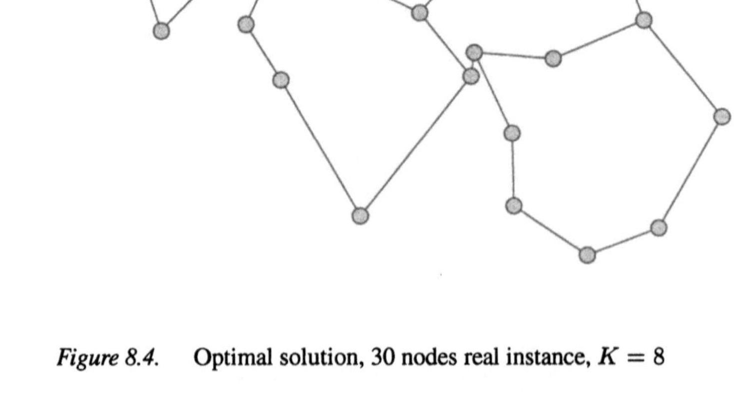

Figure 8.4. Optimal solution, 30 nodes real instance, $K = 8$

$\lvert V \rvert$	$\lvert E \rvert$	K	# ineq.	# B&B nodes	LB (root)	LB (final)	UB	Gap (%)	CPU time (hh:mm:ss)
12	39	3	179	31	740	794	794	0.0	0:00:09
12	39	4	288	105	647	681	681	0.0	0:00:18
12	39	5	73	7	598	606	606	0.0	0:00:01
12	39	6	53	5	563	568	568	0.0	0:00:01
12	39	8	39	13	521	537	537	0.0	0:00:01
12	39	10	27	7	517	521	521	0.0	0:00:01
12	39	16	24	1	496	496	496	0.0	0:00:01
17	88	3	1570	3521	1034	1100	1100	0.0	0:18:29
17	88	4	2549	5381	907	966	966	0.0	0:29:37
17	88	5	270	61	837	855	855	0.0	0:00:26
17	88	6	75	1	797	797	797	0.0	0:00:02
17	88	8	222	105	752	766	766	0.0	0:00:22
17	88	10	36	1	715	715	715	0.0	0:00:01
17	88	16	13	1	711	711	711	0.0	0:00:01
30	200	3	8481	38011	1310	1417	1431	1.0	10:00:00
30	200	4	23545	33887	1171	1227	1326	8.1	10:00:00
30	200	5	22860	35693	1078	1127	1176	4.3	10:00:00
30	200	6	2683	2721	1018	1055	1055	0.0	0:45:22
30	200	8	1957	1049	930	956	956	0.0	0:18:21
30	200	10	315	9	895	901	901	0.0	0:00:41
30	200	16	1650	781	843	861	861	0.0	0:08:31
52	622	3	4143	11115	1630	1674	1870	11.7	10:00:00
52	622	4	5129	5783	1411	1457	1663	14.1	10:00:00
52	622	5	6474	3781	1303	1354	1478	9.2	10:00:00
52	622	6	5233	4099	1256	1292	1362	5.4	10:00:00
52	622	8	5974	3049	1178	1209	1246	3.1	10:00:00
52	622	10	7535	3933	1136	1160	1208	4.1	10:00:00
52	622	16	10117	4693	1074	1091	1104	1.2	10:00:00

Table 8.5. Branch-and-Cut results, real applications, unit edge lengths

K	cut	subset	ring-cut	node-partition	cyclomatic	metric
3	232	395	135	175	220	2986
4	60	263	367	315	266	3858
5	78	409	477	323	183	5004
6	91	499	226	253	174	4290
8	115	618	127	165	136	4813
10	137	1077	182	174	207	5758
16	224	2429	216	155	153	6940

Table 8.6. **Inequalities generated for the 52 nodes real instance, unit edge lengths**

$\|V\|$	$\|E\|$	K	p/o	# ineq.	# B&B nodes	Gap (root)	Gap (final)	CPU time (hh:mm:ss)
10	25.6	3	5/5	177.8	64.2	5.7	0.0	0:00:08
10	25.6	4	5/5	93.6	35.0	4.7	0.0	0:00:03
10	25.6	5	5/5	72.0	20.6	2.5	0.0	0:00:02
10	25.6	6	5/5	48.8	14.6	3.6	0.0	0:00:01
10	25.6	8	5/5	17.0	5.4	1.0	0.0	0:00:01
10	25.6	10	5/5	3.2	1.0	0.0	0.0	0:00:01
20	74.4	3	5/4	2763.8	17192.2	8.0	1.1	2:15:54
20	68.7	4	5/5	6454.4	24128.2	8.4	0.0	2:29:58
20	68.7	5	5/5	582.8	477.4	4.4	0.0	0:02:31
20	68.7	6	5/5	355.0	169.8	3.7	0.0	0:00:58
20	68.7	8	5/5	322.8	169.4	3.4	0.0	0:00:42
20	68.7	10	5/5	228.8	37.8	2.1	0.0	0:00:19
20	68.7	16	5/5	68.6	7.0	0.7	0.0	0:00:01
30	179.4	3	5/2	4845.6	26101.4	10.8	3.9	7:54:33
30	179.4	4	5/1	15326.2	26165.4	12.2	6.3	8:46:06
30	179.4	5	5/1	15310.0	21576.6	11.1	5.3	8:03:46
30	179.4	6	5/4	7763.2	13739.8	6.4	1.0	5:03:45
30	179.4	8	5/5	2471.2	2261.4	4.1	0.0	0:55:43
30	179.4	10	5/5	2652.8	1824.6	3.1	0.0	0:43:14
30	179.4	16	5/5	2264.0	705.4	2.7	0.0	0:16:27
40	231.6	3	5/0	2237.4	8677.0	10.7	5.8	3:00:00
40	231.6	4	5/0	3790.6	7089.0	10.9	7.0	3:00:00
40	231.6	5	5/0	4203.8	6433.8	10.3	5.8	3:00:00
40	231.6	6	5/0	4863.0	7096.2	8.6	4.3	3:00:00
40	231.6	8	5/2	4336.4	4258.2	5.6	1.4	2:07:27
40	231.6	10	5/2	4573.8	3095.0	5.7	1.6	2:00:23
40	231.6	16	5/4	3850.8	1889.8	3.1	0.4	1:14:11
50	337.2	3	5/0	2089.4	4400.6	12.8	9.3	3:00:00
50	337.2	4	5/0	3123.6	3736.2	15.9	12.2	3:00:00
50	337.2	5	5/0	3680.6	3312.2	13.3	9.7	3:00:00
50	337.2	6	5/0	3871.6	3417.4	12.3	8.9	3:00:00
50	337.2	8	5/0	4227.4	2871.4	8.1	4.9	3:00:00
50	337.2	10	5/0	4315.8	2744.2	5.8	2.8	3:00:00
50	337.2	16	5/1	4448.4	2290.2	3.5	1.3	2:28:00

Table 8.7. Branch-and-Cut results, random networks, unit edge lengths

8.3 HEURISTICS FOR EUCLIDEAN EDGE LENGTHS

We now turn to results obtained with heuristics for instances with rounded Euclidean lengths. Tables 8.8 and 8.9 report the solutions obtained with the four constructive heuristics presented in the previous chapter. Computing times are not reported since they were all inferior to one second. An X denotes the case when no feasible solution was found for real instances. For randomly generated instances, means do not include instances for which one of the heuristics failed. There were 32 such instances over the 185 test problems. We also report in the tables the number of instances for which each heuristic produced the best result.

Stingy performs best, since it produced the best solution for 94 instances out of 185, and it is the only heuristic that is always able to generate a feasible solution when one exists. Therefore, we used the solution produced by stingy as initial solution for the Tabu Search heuristic.

We performed 1000 iterations of the Tabu Search heuristic. Tables 8.10 and 8.11 gather the results obtained. The solutions are always close to the optimum. The few cases where the guarantee (which is computed relatively to the best lower bound obtained with the Branch-and-Cut) is greater than 2 % correspond to problems that could not be solved by the Branch-and-Cut algorithm, meaning that it is certainly the lower bound which is not accurate.

|V|	K	ear-insert	cut-in-two	path-follow	stingy
12	200	680	**675**	783	684
12	250	**541**	558	650	**541**
12	300	558	558	576	**541**
12	350	558	**521**	**521**	**521**
12	400	567	**521**	**521**	**521**
12	450	547	521	521	**516**
12	500	**496**	**496**	**496**	**496**
17	200	923	933	985	**920**
17	250	845	849	990	**789**
17	300	823	**785**	865	789
17	350	784	746	768	**743**
17	400	767	746	**720**	735
17	450	778	706	**690**	715
17	500	758	706	**705**	715
30	150	**1304**	X	1390	1323
30	200	1194	1169	1309	**1157**
30	250	1068	1003	1231	**977**
30	300	1042	979	1218	**951**
30	350	1014	950	1057	**910**
30	400	1041	**950**	983	969
30	450	951	**912**	956	919
30	500	931	883	**879**	919
52	150	1522	1600	1618	**1367**
52	200	1380	1368	1526	**1283**
52	250	1307	1291	1429	**1185**
52	300	1345	1221	1331	**1164**
52	350	1328	**1195**	1342	1232
52	400	1275	1170	1232	**1158**
52	450	1236	1170	1287	**1130**
52	500	1257	1170	1162	**1108**
Best		3	8	7	20

Table 8.8. Constructive heuristics, real applications, Euclidean edge lengths

$\|V\|$	K	# inst.	ear-insert	cut-in-two	path-follow	stingy
10	350	1	990.0	**906.0**	1178.0	**906.0**
10	400	2	**1050.5**	1273.0	1181.5	1105.5
10	450	4	1096.8	**1090.0**	1117.0	1105.2
10	500	4	**1017.8**	1051.2	1068.5	1060.5
20	250	2	1361.5	1473.5	1515.0	**1345.0**
20	300	3	1346.3	1387.3	1374.3	**1312.3**
20	350	5	**1260.8**	1319.6	1392.8	1277.6
20	400	5	1258.6	**1207.0**	1314.6	1237.6
20	450	5	1227.6	**1162.0**	1295.2	1165.6
20	500	5	1219.0	**1109.8**	1147.4	1133.4
30	250	2	**1907.5**	2377.0	1952.0	1942.0
30	300	5	**1602.2**	1730.2	1752.0	1674.2
30	350	5	1503.6	1519.4	1697.0	**1494.2**
30	400	5	1442.8	1436.4	1607.6	**1406.2**
30	450	5	1403.4	1370.0	1490.8	**1328.8**
30	500	5	1420.2	1344.8	1323.6	**1316.0**
40	250	3	2042.3	2239.0	2144.3	**2008.3**
40	300	5	1919.6	1988.2	2013.0	**1865.4**
40	350	5	1815.4	1795.6	1917.6	**1740.0**
40	400	5	1730.4	1760.4	1879.6	**1638.8**
40	450	5	1723.0	1647.8	1798.6	**1643.6**
40	500	5	1721.2	**1573.6**	1772.8	1617.6
50	200	2	2184.5	2378.5	2410.0	**2109.5**
50	250	5	2108.6	2170.4	2326.8	**2060.6**
50	300	5	1976.4	1976.0	2260.8	**1899.4**
50	350	5	1829.0	1848.6	2164.2	**1817.0**
50	400	5	1887.8	1820.6	1996.6	**1814.2**
50	450	5	1830.8	1788.6	1945.2	**1750.6**
50	500	5	1812.6	**1671.8**	1923.8	1702.4
	Best		42	31	10	74

Table 8.9. Constructive heuristics, random networks, Euclidean edge lengths

$\|V\|$	K	stingy	tabu	best iteration	CPU time	guarantee (%)
12	200	684	622	54	37	0.0
12	250	541	541	0	30	0.0
12	300	541	541	0	24	0.0
12	350	521	521	0	24	0.0
12	400	521	521	0	19	0.0
12	450	516	516	0	16	0.0
12	500	496	496	0	15	0.0
17	200	920	834	5	129	0.0
17	250	789	789	0	104	0.0
17	300	789	726	2	77	0.0
17	350	743	725	6	66	0.0
17	400	735	720	184	63	0.0
17	450	715	689	21	55	0.0
17	500	715	689	14	52	0.0
30	150	1323	1288	145	982	9.2
30	200	1157	1065	381	559	0.0
30	250	977	938	823	454	0.1
30	300	951	898	223	350	0.0
30	350	910	874	109	360	0.0
30	400	969	870	306	276	0.2
30	450	919	869	348	266	0.1
30	500	919	844	205	263	0.0
52	150	1367	1339	47	4090	7.2
52	200	1283	1225	16	2538	5.2
52	250	1185	1165	513	2280	3.6
52	300	1164	1152	8	1677	5.1
52	350	1232	1104	7	1909	2.0
52	400	1158	1081	336	2173	1.4
52	450	1130	1068	37	1627	1.4
52	500	1108	1068	46	1578	1.4

Table 8.10. Tabu Search, real applications, Euclidean edge lengths

$\lvert V\rvert$	K	# inst.	stingy	tabu	best iteration	CPU time	guarantee (%)
10	300	2	1439.0	1330.0	2.0	10.5	0.0
10	350	4	1177.0	1128.0	13.5	15.8	0.0
10	400	5	1159.4	1069.4	30.8	19.6	0.0
10	450	5	1096.6	1007.0	91.2	20.0	0.0
10	500	5	1060.8	977.8	112.8	19.0	0.0
20	200	4	1673.2	1638.8	76.2	168.2	3.9
20	250	4	1365.8	1305.2	155.8	175.8	0.2
20	300	5	1324.4	1225.0	84.8	184.2	0.3
20	350	5	1277.6	1158.4	67.0	183.2	0.0
20	400	5	1237.6	1126.4	330.2	180.6	0.2
20	450	5	1165.6	1093.2	283.8	157.6	0.4
20	500	5	1133.4	1055.6	202.8	146.2	0.4
30	200	3	1952.3	1867.7	167.7	848.3	7.3
30	250	4	1785.5	1607.0	331.0	823.5	8.9
30	300	5	1674.2	1475.2	525.8	744.0	5.3
30	350	5	1494.2	1339.2	416.4	575.2	0.8
30	400	5	1406.2	1288.6	456.2	454.4	1.4
30	450	5	1328.8	1248.4	230.0	425.4	0.5
30	500	5	1316.0	1231.0	493.8	397.8	0.9
40	200	4	2219.0	2084.5	172.0	1542.5	9.3
40	250	5	2002.6	1859.6	290.2	1815.8	10.1
40	300	5	1865.4	1739.2	29.0	1594.0	9.7
40	350	5	1740.0	1610.0	365.4	1379.8	5.7
40	400	5	1638.8	1559.0	247.8	1235.6	5.4
40	450	5	1643.6	1510.8	402.4	1150.6	4.2
40	500	5	1617.6	1463.6	535.0	1035.2	2.0
50	150	1	2367.0	2250.0	295.0	4358.0	3.9
50	200	4	2152.5	2017.8	348.0	3373.8	10.9
50	250	5	2060.6	1926.6	517.4	3384.2	10.9
50	300	5	1899.4	1752.8	176.2	2835.4	6.7
50	350	5	1817.0	1672.0	271.6	2741.2	4.3
50	400	5	1814.2	1649.8	324.6	2454.0	4.9
50	450	5	1750.6	1633.4	243.6	2352.6	5.4
50	500	5	1702.4	1602.4	269.6	2226.8	4.6

Table 8.11. Tabu Search, random networks, Euclidean edge lengths

8.4 HEURISTICS FOR UNIT EDGE LENGTHS

We present in Tables 8.12 and 8.13 results obtained with the stingy and the Tabu Search heuristic for problems with unit edge lengths. The three other heuristics were first developed for the case where edge costs are equal to their lengths, and since these heuristics turned out to be ineffective compared to the stingy, we did not devote time to adapting our code.

Although the quality of the solutions obtained is good for problems of small size or sufficiently large values of the bound ($K \geq 6$), for large problems with a small bound, solutions obtained are of bad quality (even compared to the upper bound found by the Branch-and-Cut algorithm) and computing times become very high. This suggests the need for a more sophisticated search in these cases.

$\|V\|$	K	stingy	tabu	best iteration	CPU time	guarantee (%)
12	3	829	794	5	46	0.0
12	4	693	681	92	28	0.0
12	5	633	606	1	18	0.0
12	6	594	568	32	13	0.0
12	8	541	537	1	9	0.0
12	10	521	521	0	7	0.0
12	16	496	496	0	5	0.0
17	3	1214	1111	48	200	1.0
17	4	1021	971	2	169	0.5
17	5	892	855	507	107	0.0
17	6	852	797	855	67	0.0
17	8	789	766	105	46	0.0
17	10	715	715	0	34	0.0
17	16	715	714	4	21	0.4
30	3	1686	1506	10	1500	6.3
30	4	1392	1306	445	867	6.4
30	5	1266	1165	939	745	3.4
30	6	1191	1067	778	484	1.1
30	8	1023	957	68	259	0.1
30	10	951	901	523	170	0.0
30	16	919	862	967	105	0.1
52	3	2039	1919	6	16911	14.6
52	4	1730	1685	73	14753	15.6
52	5	1556	1482	78	10029	9.5
52	6	1445	1365	896	5362	5.7
52	8	1306	1249	218	2888	3.3
52	10	1343	1201	422	2020	3.5
52	16	1251	1134	715	799	3.9

Table 8.12. Tabu Search, real applications, unit edge lengths

$\|V\|$	K	# inst.	stingy	tabu	best iteration	CPU time	guarantee (%)
10	3	5	1375.4	1321.6	13.2	18.0	0.0
10	4	5	1138.6	1088.2	62.0	10.6	0.0
10	5	5	1064.4	992.6	5.4	7.4	0.0
10	6	5	1005.4	938.8	5.0	5.2	0.0
10	8	5	963.0	893.6	4.6	3.6	0.0
10	10	5	937.0	839.6	6.4	2.6	0.1
20	3	5	1755.8	1669.8	189.8	257.6	2.5
20	4	5	1460.6	1426.6	209.2	171.2	1.7
20	5	5	1342.8	1268.4	164.6	102.8	2.0
20	6	5	1310.4	1208.8	335.0	74.2	3.4
20	8	5	1169.4	1105.0	153.0	47.0	0.2
20	10	5	1113.4	1059.8	144.8	32.4	0.2
20	16	5	1073.6	1002.2	402.8	20.6	0.1
30	3	5	2195.2	2080.0	124.2	1321.8	5.6
30	4	5	1885.8	1780.0	301.0	987.0	6.7
30	5	5	1728.6	1621.6	545.6	757.6	5.3
30	6	5	1597.0	1503.0	484.4	492.8	2.5
30	8	5	1455.0	1342.0	382.6	293.6	0.6
30	10	5	1386.4	1291.2	226.2	181.8	1.9
30	16	5	1303.0	1190.0	288.0	105.2	0.4
40	3	5	2576.2	2453.0	18.8	3207.4	7.3
40	4	5	2271.0	2118.4	529.6	1899.2	6.8
40	5	5	2059.6	1946.0	810.0	1257.8	5.6
40	6	5	1938.6	1820.0	294.2	1064.4	4.2
40	8	5	1798.0	1654.6	247.8	579.0	2.0
40	10	5	1700.4	1563.6	356.8	416.0	2.4
40	16	5	1529.8	1435.0	214.8	209.4	1.4
50	3	5	2889.4	2797.4	211.2	10101.8	16.7
50	4	5	2443.4	2336.2	192.4	5391.0	12.1
50	5	5	2245.2	2126.4	332.8	3466.0	9.6
50	6	5	2157.0	2005.8	509.4	2565.8	8.7
50	8	5	1999.4	1819.4	347.8	1167.8	5.2
50	10	5	1858.2	1713.6	356.0	939.8	3.1
50	16	5	1738.4	1615.2	421.2	447.8	2.9

Table 8.13. Tabu Search, random networks, unit edge lengths

Chapter 9

CONCLUSION

In this work, we studied a problem arising from the need to design survivable telecommunication networks. Designing a two-connected network at minimum cost is a problem that was widely studied in the literature, and efficient methods for solving it were already available.

Unfortunately, it turned out that the optimal solution of this problem on the set of 52 zonal centers of the Belgian backbone network is a Hamiltonian cycle. Hence, any edge failure would require to reroute the flow that passed through that edge, using all the edges of the network, an obviously undesirable feature. This led us to examine a new model for limiting the region of influence of the traffic which is necessary to reroute : in addition to two-connectivity, we required that each edge belongs to at least one cycle whose length is bounded by a given constant. This problem is called the *Two-Connected Network with Bounded Rings problem.*

Results presented in this dissertation are, to our knowledge, the first attempt to solve the problem. The problem was formulated as an integer linear program, and a deep polyhedral study was done, leading to a Branch-and-Cut algorithm. Additional polyhedral results were obtained for the (more difficult) case of unit edge lengths. Finally, since we could only solve small problems to optimality with the Branch-and-Cut algorithm, we developed a series of heuristics, among which a

Tabu Search one which turned out to be quite effective. However, problems with unit edge lengths and a small value of the bound (3 to 5) remain difficult to deal with and should probably need specific methods.

Appendix A
Detailed Computational Results

A.1 DESCRIPTION OF THE TEST PROBLEMS

In order to avoid useless redundancy in the tables reporting numerical results, we present here the characteristics of the different test problems that were used.

Two classes of problems are solved, defined by different edge lengths. In the first class, edge lengths are equal to the rounded Euclidean distances. The second class is composed of problems with unit edge lengths. In all cases, we consider costs equal to the rounded Euclidean distance.

Tests were made for different values of the bound, for instances coming from real applications, with 12, 17, 30 and 52 nodes, and for random problems with nodes uniformly generated in a square of size 250×250. Random problems with 10 to 50 nodes were generated, and we tested five instances of each size.

Data on the randomly generated test problems are available at the Web page `http://smg.ulb.ac.be/~bfortz/2cnbm/data.html`.

When edge lengths are equal to the rounded Euclidean distances, we apply the preprocessing presented in Section 6.9, allowing to reduce the number of edges. Table A.1 reports the size of the problems in function of the bound for real applications, while Tables A.2 to A.6 report these sizes for randomly generated problems. We also mention in these tables values of the bound leading to infeasible problems.

In the case of unit lengths, we already mentioned that the preprocessing does not allow to remove edges, and we are therefore left with a complete graph. To overcome this drawback, it seems quite natural to avoid edges with a too large cost. We therefore decided to suppress from the graph all edges having a cost greater than 80. If the resulting graph is infeasible, we increase this maximal cost by 20 until the graph becomes feasible. Table A.7 reports the size of the resulting problems and the maximal cost used.

Graph	K	# nodes	# edges	
B12	150	12	31	infeasible
B12	200	12	49	
B12	250	12	58	
B12	300	12	63	
B12	350	12	65	
B12	400	12	66	
B12	450	12	66	
B12	500	12	66	
B17	150	17	55	infeasible
B17	200	17	88	
B17	250	17	109	
B17	300	17	119	
B17	350	17	126	
B17	400	17	133	
B17	450	17	135	
B17	500	17	136	
B30	150	30	174	
B30	200	30	266	
B30	250	30	328	
B30	300	30	372	
B30	350	30	406	
B30	400	30	422	
B30	450	30	432	
B30	500	30	435	
B52	150	52	559	
B52	200	52	821	
B52	250	52	1040	
B52	300	52	1149	
B52	350	52	1230	
B52	400	52	1287	
B52	450	52	1313	
B52	500	52	1324	

Table A.1. Real applications, Euclidean edge lengths

Graph	K	# nodes	# edges	
R10-1	150	10	6	infeasible
R10-1	200	10	13	infeasible
R10-1	250	10	18	infeasible
R10-1	300	10	21	infeasible
R10-1	350	10	27	
R10-1	400	10	36	
R10-1	450	10	41	
R10-1	500	10	45	
R10-2	150	10	8	infeasible
R10-2	200	10	15	infeasible
R10-2	250	10	15	infeasible
R10-2	300	10	18	infeasible
R10-2	350	10	22	infeasible
R10-2	400	10	30	
R10-2	450	10	39	
R10-2	500	10	43	
R10-3	150	10	0	infeasible
R10-3	200	10	6	infeasible
R10-3	250	10	17	infeasible
R10-3	300	10	25	
R10-3	350	10	32	
R10-3	400	10	38	
R10-3	450	10	39	
R10-3	500	10	44	
R10-4	150	10	0	infeasible
R10-4	200	10	3	infeasible
R10-4	250	10	12	infeasible
R10-4	300	10	21	
R10-4	350	10	31	
R10-4	400	10	34	
R10-4	450	10	41	
R10-4	500	10	42	
R10-5	150	10	5	infeasible
R10-5	200	10	9	infeasible
R10-5	250	10	11	infeasible
R10-5	300	10	14	infeasible
R10-5	350	10	25	
R10-5	400	10	35	
R10-5	450	10	40	
R10-5	500	10	42	

Table A.2. Random networks (10 nodes), Euclidean edge lengths

Graph	K	# nodes	# edges	
R20-1	150	20	23	infeasible
R20-1	200	20	52	
R20-1	250	20	83	
R20-1	300	20	112	
R20-1	350	20	144	
R20-1	400	20	160	
R20-1	450	20	183	
R20-1	500	20	187	
R20-2	150	20	46	infeasible
R20-2	200	20	82	
R20-2	250	20	117	
R20-2	300	20	146	
R20-2	350	20	170	
R20-2	400	20	181	
R20-2	450	20	189	
R20-2	500	20	190	
R20-3	150	20	36	infeasible
R20-3	200	20	70	
R20-3	250	20	89	
R20-3	300	20	118	
R20-3	350	20	151	
R20-3	400	20	177	
R20-3	450	20	185	
R20-3	500	20	188	
R20-4	150	20	26	infeasible
R20-4	200	20	59	
R20-4	250	20	97	
R20-4	300	20	124	
R20-4	350	20	146	
R20-4	400	20	169	
R20-4	450	20	177	
R20-4	500	20	186	
R20-5	150	20	34	infeasible
R20-5	200	20	62	infeasible
R20-5	250	20	81	infeasible
R20-5	300	20	109	
R20-5	350	20	137	
R20-5	400	20	164	
R20-5	450	20	178	
R20-5	500	20	185	

Table A.3. Random networks (20 nodes), Euclidean edge lengths

Graph	K	# nodes	# edges	
R30-1	150	30	70	infeasible
R30-1	200	30	133	
R30-1	250	30	207	
R30-1	300	30	276	
R30-1	350	30	332	
R30-1	400	30	373	
R30-1	450	30	409	
R30-1	500	30	423	
R30-2	150	30	117	infeasible
R30-2	200	30	182	infeasible
R30-2	250	30	235	infeasible
R30-2	300	30	284	
R30-2	350	30	342	
R30-2	400	30	379	
R30-2	450	30	415	
R30-2	500	30	427	
R30-3	150	30	107	infeasible
R30-3	200	30	157	infeasible
R30-3	250	30	241	
R30-3	300	30	306	
R30-3	350	30	362	
R30-3	400	30	394	
R30-3	450	30	422	
R30-3	500	30	431	
R30-4	150	30	98	infeasible
R30-4	200	30	170	
R30-4	250	30	259	
R30-4	300	30	318	
R30-4	350	30	374	
R30-4	400	30	410	
R30-4	450	30	427	
R30-4	500	30	434	
R30-5	150	30	63	infeasible
R30-5	200	30	134	
R30-5	250	30	198	
R30-5	300	30	260	
R30-5	350	30	319	
R30-5	400	30	352	
R30-5	450	30	401	
R30-5	500	30	419	

Table A.4. Random networks (30 nodes), Euclidean edge lengths

Graph	K	# nodes	# edges	
R40-1	150	40	127	infeasible
R40-1	200	40	237	
R40-1	250	40	348	
R40-1	300	40	458	
R40-1	350	40	556	
R40-1	400	40	628	
R40-1	450	40	715	
R40-1	500	40	752	
R40-2	150	40	145	infeasible
R40-2	200	40	241	infeasible
R40-2	250	40	375	
R40-2	300	40	464	
R40-2	350	40	549	
R40-2	400	40	638	
R40-2	450	40	719	
R40-2	500	40	752	
R40-3	150	40	107	infeasible
R40-3	200	40	220	
R40-3	250	40	328	
R40-3	300	40	421	
R40-3	350	40	521	
R40-3	400	40	621	
R40-3	450	40	694	
R40-3	500	40	738	
R40-4	150	40	138	infeasible
R40-4	200	40	268	
R40-4	250	40	382	
R40-4	300	40	478	
R40-4	350	40	569	
R40-4	400	40	644	
R40-4	450	40	714	
R40-4	500	40	749	
R40-5	150	40	203	infeasible
R40-5	200	40	307	
R40-5	250	40	411	
R40-5	300	40	512	
R40-5	350	40	613	
R40-5	400	40	697	
R40-5	450	40	757	
R40-5	500	40	777	

Table A.5. Random networks (40 nodes), Euclidean edge lengths

Graph	K	# nodes	# edges	
R50-1	150	50	243	
R50-1	200	50	394	
R50-1	250	50	550	
R50-1	300	50	704	
R50-1	350	50	874	
R50-1	400	50	1049	
R50-1	450	50	1151	
R50-1	500	50	1202	
R50-2	150	50	228	infeasible
R50-2	200	50	392	infeasible
R50-2	250	50	553	
R50-2	300	50	712	
R50-2	350	50	861	
R50-2	400	50	995	
R50-2	450	50	1106	
R50-2	500	50	1182	
R50-3	150	50	255	infeasible
R50-3	200	50	440	
R50-3	250	50	621	
R50-3	300	50	826	
R50-3	350	50	966	
R50-3	400	50	1075	
R50-3	450	50	1152	
R50-3	500	50	1197	
R50-4	150	50	272	infeasible
R50-4	200	50	435	
R50-4	250	50	608	
R50-4	300	50	763	
R50-4	350	50	917	
R50-4	400	50	1034	
R50-4	450	50	1112	
R50-4	500	50	1187	
R50-5	150	50	293	infeasible
R50-5	200	50	500	
R50-5	250	50	703	
R50-5	300	50	872	
R50-5	350	50	1012	
R50-5	400	50	1123	
R50-5	450	50	1184	
R50-5	500	50	1209	

Table A.6. Random networks (50 nodes), Euclidean edge lengths

Graph	max. cost	# nodes	# edges	
B12	80	12	39	
B17	100	17	88	
B30	80	30	200	
B52	80	52	622	
R10-1	140	10	23	
R10-2	200	10	31	
R10-3	120	10	22	
R10-4	160	10	29	
R10-5	160	10	23	
R20-1	100	20	60	$(K = 3)$
R20-1	80	20	40	$(K \geq 4)$
R20-2	100	20	85	
R20-3	100	20	73	
R20-4	100	20	69	
R20-5	120	20	85	
R30-1	80	30	92	
R30-2	160	30	314	
R30-3	100	30	163	
R30-4	100	30	183	
R30-5	100	30	145	
R40-1	100	40	248	
R40-2	120	40	358	
R40-3	80	40	140	
R40-4	80	40	182	
R40-5	80	40	230	
R50-1	80	50	281	
R50-2	80	50	280	
R50-3	80	50	304	
R50-4	80	50	316	
R50-5	100	50	505	

Table A.7. Instances with unit edge lengths

A.2 BRANCH-AND-CUT FOR EUCLIDEAN EDGE LENGTHS

Graph	K	# ineq.	# B&B nodes	LB (root)	LB (final)	UB	Gap (%)	CPU time (hh:mm:ss)
B12	200	220	35	577	622	622	0.0	0:00:09
B12	250	64	7	530	541	541	0.0	0:00:01
B12	300	125	35	514	541	541	0.0	0:00:27
B12	350	65	5	509	521	521	0.0	0:00:01
B12	400	92	13	503	521	521	0.0	0:00:02
B12	450	79	15	499	516	516	0.0	0:00:12
B12	500	27	1	496	496	496	0.0	0:00:01
B17	200	296	39	802	834	834	0.0	0:00:22
B17	250	379	103	743	789	789	0.0	0:00:47
B17	300	144	23	711	726	726	0.0	0:00:08
B17	350	223	45	701	725	725	0.0	0:00:14
B17	400	336	139	688	720	720	0.0	0:00:36
B17	450	69	7	680	689	689	0.0	0:00:03
B17	500	93	11	675	689	689	0.0	0:00:05
B30	150	34851	44411	1057	1180	1258	6.6	10:00:00
B30	200	25477	32175	973	1065	1065	0.0	8:26:55
B30	250	587	155	898	937	937	0.0	0:06:27
B30	300	1090	815	875	898	898	0.0	0:15:53
B30	350	390	115	858	874	874	0.0	0:03:51
B30	400	559	365	849	868	868	0.0	0:07:04
B30	450	1622	1313	843	868	868	0.0	0:26:19
B30	500	273	47	837	844	844	0.0	0:01:45
B52	150	7877	12127	1198	1249	1339	7.2	10:00:00
B52	200	6514	7475	1132	1165	1239	6.4	10:00:00
B52	250	5161	4845	1097	1125	1184	5.2	10:00:00
B52	300	5220	3627	1073	1096	1138	3.8	10:00:00
B52	350	4486	3727	1061	1082	1085	0.3	10:00:00
B52	400	2828	1407	1048	1066	1066	0.0	4:32:55
B52	450	926	195	1042	1053	1053	0.0	0:54:33
B52	500	1558	417	1039	1053	1053	0.0	1:49:12

Table A.8. Branch-and-Cut results, real applications, Euclidean edge lengths

Graph	K	# ineq.	# B&B nodes	LB (root)	LB (final)	UB	Gap (%)	CPU time (hh:mm:ss)
R10-1	350	61	3	905	906	906	0.0	0:00:01
R10-1	400	141	11	826	898	898	0.0	0:00:02
R10-1	450	427	69	790	896	896	0.0	0:00:13
R10-1	500	283	61	774	854	854	0.0	0:00:08
R10-2	400	143	9	1075	1140	1140	0.0	0:00:03
R10-2	450	264	19	975	1062	1062	0.0	0:00:04
R10-2	500	202	25	943	1031	1031	0.0	0:00:04
R10-3	300	107	19	1098	1269	1269	0.0	0:00:03
R10-3	350	113	17	963	1018	1018	0.0	0:00:03
R10-3	400	92	7	936	954	954	0.0	0:00:01
R10-3	450	149	19	904	952	952	0.0	0:00:03
R10-3	500	187	41	875	952	952	0.0	0:00:06
R10-4	300	81	1	1391	1391	1391	0.0	0:00:01
R10-4	350	211	35	1068	1205	1205	0.0	0:00:06
R10-4	400	283	55	994	1117	1117	0.0	0:00:09
R10-4	450	131	9	932	980	980	0.0	0:00:02
R10-4	500	152	33	911	980	980	0.0	0:00:04
R10-5	350	115	5	1254	1383	1383	0.0	0:00:01
R10-5	400	326	59	1016	1238	1238	0.0	0:00:12
R10-5	450	485	127	965	1143	1143	0.0	0:00:23
R10-5	500	332	65	937	1072	1072	0.0	0:00:13

Table A.9. Branch-and-Cut results, random networks (10 nodes), Euclidean edge lengths

Graph	K	# ineq.	# B&B nodes	LB (root)	LB (final)	UB	Gap (%)	CPU time (hh:mm:ss)
R20-1	200	324	31	1393	1577	1577	0.0	0:00:20
R20-1	250	3332	4851	1208	1445	1445	0.0	0:31:08
R20-1	300	47471	70667	1154	1376	1376	0.0	7:29:02
R20-1	350	3171	5581	1113	1253	1253	0.0	0:36:57
R20-1	400	940	677	1083	1183	1183	0.0	0:07:02
R20-1	450	1050	579	1067	1144	1144	0.0	0:07:30
R20-1	500	939	319	1045	1111	1111	0.0	0:05:16
R20-2	200	702	335	1210	1325	1325	0.0	0:03:19
R20-2	250	2813	2957	974	1094	1094	0.0	0:22:10
R20-2	300	517	143	938	984	984	0.0	0:01:36
R20-2	350	986	653	906	953	953	0.0	0:05:57
R20-2	400	995	575	891	940	940	0.0	0:05:33
R20-2	450	1008	377	874	919	919	0.0	0:04:24
R20-2	500	650	151	864	900	900	0.0	0:02:21
R20-3	200	344	41	1373	1449	1449	0.0	0:00:35
R20-3	250	700	301	1075	1218	1218	0.0	0:02:05
R20-3	300	377	87	1029	1134	1134	0.0	0:00:57
R20-3	350	719	291	994	1100	1100	0.0	0:03:46
R20-3	400	1576	1005	975	1100	1100	0.0	0:12:22
R20-3	450	11295	12413	954	1100	1100	0.0	1:54:39
R20-3	500	752	243	932	1011	1011	0.0	0:03:33
R20-4	200	394	109	1862	1958	1958	0.0	0:01:29
R20-4	250	924	543	1316	1454	1454	0.0	0:04:30
R20-4	300	861	419	1203	1286	1286	0.0	0:03:31
R20-4	350	661	207	1158	1235	1235	0.0	0:02:24
R20-4	400	681	245	1132	1190	1190	0.0	0:02:46
R20-4	450	562	221	1114	1164	1164	0.0	0:02:14
R20-4	500	535	195	1101	1149	1149	0.0	0:02:02
R20-5	300	1070	475	1242	1324	1324	0.0	0:04:52
R20-5	350	1001	749	1127	1251	1251	0.0	0:06:07
R20-5	400	1426	927	1079	1211	1211	0.0	0:09:40
R20-5	450	428	83	1035	1119	1119	0.0	0:01:43
R20-5	500	519	133	1010	1082	1082	0.0	0:02:05

Table A.10. Branch-and-Cut results, random networks (20 nodes), Euclidean edge lengths

Graph	K	# ineq.	# B&B nodes	LB (root)	LB (final)	UB	Gap (%)	CPU time (hh:mm:ss)
R30-1	200	22621	40791	1480	1726	1726	0.0	8:45:52
R30-1	250	31787	40315	1361	1477	1613	9.2	10:00:00
R30-1	300	27341	35479	1293	1413	1474	4.3	10:00:00
R30-1	350	21888	28759	1249	1328	1393	4.9	10:00:00
R30-1	400	19039	22929	1202	1295	1324	2.2	10:00:00
R30-1	450	1844	981	1165	1240	1240	0.0	0:35:04
R30-1	500	2179	1515	1145	1213	1213	0.0	1:00:42
R30-2	300	6348	7381	1392	1477	1477	0.0	2:42:46
R30-2	350	3215	2923	1306	1381	1381	0.0	1:17:21
R30-2	400	1151	573	1268	1319	1319	0.0	0:20:06
R30-2	450	3383	3651	1231	1319	1319	0.0	1:57:16
R30-2	500	2058	1575	1209	1295	1295	0.0	0:49:26
R30-3	250	20269	27965	1311	1411	1587	12.5	10:00:00
R30-3	300	5710	8913	1216	1325	1325	0.0	3:18:46
R30-3	350	1314	701	1179	1239	1239	0.0	0:21:49
R30-3	400	1240	831	1156	1198	1198	0.0	0:28:47
R30-3	450	567	107	1112	1152	1152	0.0	0:07:34
R30-3	500	590	151	1099	1128	1128	0.0	0:07:22
R30-4	200	25457	35769	1372	1448	1549	7.0	10:00:00
R30-4	250	21892	26865	1149	1250	1327	6.2	10:00:00
R30-4	300	23444	20399	1105	1164	1301	11.8	10:00:00
R30-4	350	4757	5233	1065	1134	1134	0.0	2:48:12
R30-4	400	473	91	1023	1050	1050	0.0	0:06:17
R30-4	450	742	323	1012	1044	1044	0.0	0:14:35
R30-4	500	1372	715	1001	1044	1044	0.0	0:27:58
R30-5	200	4316	5863	1814	2056	2056	0.0	1:19:28
R30-5	250	30432	31799	1611	1759	1914	8.8	10:00:00
R30-5	300	26361	27657	1526	1635	1780	8.9	10:00:00
R30-5	350	5082	6847	1446	1562	1562	0.0	2:07:52
R30-5	400	3017	3429	1411	1493	1493	0.0	1:09:53
R30-5	450	2680	2965	1383	1452	1452	0.0	1:08:58
R30-5	500	2216	1899	1367	1424	1424	0.0	0:52:43

Table A.11. Branch-and-Cut results, random networks (30 nodes), Euclidean edge lengths

Graph	K	# ineq.	# B&B nodes	LB (root)	LB (final)	UB	Gap (%)	CPU time (hh:mm:ss)
R40-1	200	5417	5967	1935	2067	2232	8.0	3:00:00
R40-1	250	4552	4365	1684	1800	2020	12.2	3:00:00
R40-1	300	3434	2739	1590	1687	1947	15.4	3:00:00
R40-1	350	3165	2597	1537	1616	1691	4.6	3:00:00
R40-1	400	3379	2745	1495	1558	1603	2.9	3:00:00
R40-1	450	1612	487	1477	1533	1533	0.0	1:47:47
R40-1	500	2506	1175	1455	1520	1569	3.2	3:00:00
R40-2	250	5219	5527	1545	1652	1765	6.8	3:00:00
R40-2	300	3793	3931	1478	1558	1621	4.0	3:00:00
R40-2	350	3198	3067	1430	1496	1514	1.2	3:00:00
R40-2	400	3202	3213	1398	1459	1473	1.0	3:00:00
R40-2	450	2525	1735	1381	1434	1455	1.5	3:00:00
R40-2	500	2637	1475	1365	1416	1422	0.4	3:00:00
R40-3	200	5700	5831	1916	2031	2317	14.1	3:00:00
R40-3	250	4496	5743	1737	1821	2067	13.5	3:00:00
R40-3	300	4193	4047	1626	1688	1815	7.5	3:00:00
R40-3	350	3471	3765	1567	1620	1654	2.1	3:00:00
R40-3	400	3703	3769	1535	1582	1615	2.1	3:00:00
R40-3	450	2760	2619	1510	1561	1586	1.6	3:00:00
R40-3	500	2670	2267	1494	1539	1560	1.4	3:00:00
R40-4	200	5391	5607	1740	1894	2079	9.8	3:00:00
R40-4	250	5320	4641	1606	1718	1814	5.6	3:00:00
R40-4	300	4014	3995	1526	1610	1678	4.2	3:00:00
R40-4	350	4104	4149	1485	1551	1583	2.1	3:00:00
R40-4	400	3949	3287	1450	1503	1561	3.9	3:00:00
R40-4	450	3510	2821	1427	1476	1523	3.2	3:00:00
R40-4	500	2817	2311	1415	1458	1475	1.2	3:00:00
R40-5	200	3836	4411	1502	1626	1718	5.7	3:00:00
R40-5	250	3347	3109	1344	1455	1676	15.2	3:00:00
R40-5	300	1891	1883	1305	1393	1599	14.8	3:00:00
R40-5	350	1599	693	1272	1356	1600	18.0	3:00:00
R40-5	400	2127	1353	1244	1315	1358	3.3	3:00:00
R40-5	450	1990	539	1211	1266	1402	10.7	3:00:00
R40-5	500	2061	823	1186	1246	1291	3.6	3:00:00

Table A.12. Branch-and-Cut results, random networks (40 nodes), Euclidean edge lengths

Graph	K	# ineq.	# B&B nodes	LB (root)	LB (final)	UB	Gap (%)	CPU time (hh:mm:ss)
R50-1	150	3566	4917	2056	2165	2250	3.9	3:00:00
R50-1	200	3283	2931	1670	1776	1968	10.8	3:00:00
R50-1	250	1881	1495	1570	1659	1884	13.6	3:00:00
R50-1	300	1781	367	1514	1569	1685	7.4	3:00:00
R50-1	350	1751	369	1469	1521	1605	5.5	3:00:00
R50-1	400	1822	449	1444	1485	1538	3.6	3:00:00
R50-1	450	1465	375	1430	1462	1527	4.4	3:00:00
R50-1	500	1448	231	1413	1444	1503	4.1	3:00:00
R50-2	250	1914	1073	1798	1884	2162	14.8	3:00:00
R50-2	300	1775	1139	1701	1772	1858	4.9	3:00:00
R50-2	350	1278	295	1667	1711	1841	7.6	3:00:00
R50-2	400	1603	1065	1643	1686	1781	5.6	3:00:00
R50-2	450	1273	455	1612	1654	1728	4.5	3:00:00
R50-2	500	1479	225	1601	1632	1725	5.7	3:00:00
R50-3	200	2526	2273	1814	1877	2053	9.4	3:00:00
R50-3	250	1569	1661	1711	1777	1929	8.6	3:00:00
R50-3	300	1187	597	1648	1697	1868	10.1	3:00:00
R50-3	350	1386	601	1606	1654	1750	5.8	3:00:00
R50-3	400	1913	763	1582	1621	1697	4.7	3:00:00
R50-3	450	1462	241	1563	1597	1740	9.0	3:00:00
R50-3	500	1638	193	1549	1582	1639	3.6	3:00:00
R50-4	200	3814	3161	1756	1852	2090	12.9	3:00:00
R50-4	250	2775	2805	1652	1709	1824	6.7	3:00:00
R50-4	300	2522	2271	1590	1634	1683	3.0	3:00:00
R50-4	350	2137	739	1540	1582	1624	2.7	3:00:00
R50-4	400	1958	1145	1518	1557	1619	4.0	3:00:00
R50-4	450	1824	837	1498	1535	1600	4.2	3:00:00
R50-4	500	1752	747	1488	1520	1585	4.3	3:00:00
R50-5	200	1976	1107	1692	1777	1960	10.3	3:00:00
R50-5	250	1746	623	1594	1650	1835	11.2	3:00:00
R50-5	300	1788	685	1507	1543	1615	4.7	3:00:00
R50-5	350	1788	685	1507	1543	1615	4.7	3:00:00
R50-5	400	1568	693	1482	1515	1596	5.3	3:00:00
R50-5	450	2042	835	1467	1501	1549	3.2	3:00:00
R50-5	500	1412	325	1456	1478	1544	4.5	3:00:00

Table A.13. Branch-and-Cut results, random networks (50 nodes), Euclidean edge lengths

A.3 BRANCH-AND-CUT FOR UNIT EDGE LENGTHS

Graph	K	# ineq.	# B&B nodes	LB (root)	LB (final)	UB	Gap (%)	CPU time (hh:mm:ss)
B12	3	179	31	740	794	794	0.0	0:00:09
B12	4	288	105	647	681	681	0.0	0:00:18
B12	5	73	7	598	606	606	0.0	0:00:01
B12	6	53	5	563	568	568	0.0	0:00:01
B12	8	39	13	521	537	537	0.0	0:00:01
B12	10	27	7	517	521	521	0.0	0:00:01
B12	16	24	1	496	496	496	0.0	0:00:01
B17	3	1570	3521	1034	1100	1100	0.0	0:18:29
B17	4	2549	5381	907	966	966	0.0	0:29:37
B17	5	270	61	837	855	855	0.0	0:00:26
B17	6	75	1	797	797	797	0.0	0:00:02
B17	8	222	105	752	766	766	0.0	0:00:22
B17	10	36	1	715	715	715	0.0	0:00:01
B17	16	13	1	711	711	711	0.0	0:00:01
B30	3	8481	38011	1310	1417	1431	1.0	10:00:00
B30	4	23545	33887	1171	1227	1326	8.1	10:00:00
B30	5	22860	35693	1078	1127	1176	4.3	10:00:00
B30	6	2683	2721	1018	1055	1055	0.0	0:45:22
B30	8	1957	1049	930	956	956	0.0	0:18:21
B30	10	315	9	895	901	901	0.0	0:00:41
B30	16	1650	781	843	861	861	0.0	0:08:31
B52	3	4143	11115	1630	1674	1870	11.7	10:00:00
B52	4	5129	5783	1411	1457	1663	14.1	10:00:00
B52	5	6474	3781	1303	1354	1478	9.2	10:00:00
B52	6	5233	4099	1256	1292	1362	5.4	10:00:00
B52	8	5974	3049	1178	1209	1246	3.1	10:00:00
B52	10	7535	3933	1136	1160	1208	4.1	10:00:00
B52	16	10117	4693	1074	1091	1104	1.2	10:00:00

Table A.14. Branch-and-Cut results, real applications, unit edge lengths

Graph	K	# ineq.	# B&B nodes	LB (root)	LB (final)	UB	Gap (%)	CPU time (hh:mm:ss)
R10-1	3	101	19	1139	1207	1207	0.0	0:00:02
R10-1	4	62	11	964	998	998	0.0	0:00:01
R10-1	5	26	3	896	901	901	0.0	0:00:01
R10-1	6	14	7	846	867	867	0.0	0:00:01
R10-1	8	12	1	825	825	825	0.0	0:00:01
R10-1	10	5	1	792	792	792	0.0	0:00:01
R10-2	3	451	237	1267	1359	1359	0.0	0:00:32
R10-2	4	46	13	1065	1109	1109	0.0	0:00:01
R10-2	5	86	55	1035	1062	1062	0.0	0:00:04
R10-2	6	48	17	1011	1022	1022	0.0	0:00:01
R10-2	8	8	1	1007	1007	1007	0.0	0:00:01
R10-2	10	6	1	993	993	993	0.0	0:00:01
R10-3	3	82	11	1227	1276	1276	0.0	0:00:01
R10-3	4	67	9	1027	1057	1057	0.0	0:00:01
R10-3	5	43	1	954	954	954	0.0	0:00:01
R10-3	6	97	29	876	931	931	0.0	0:00:02
R10-3	8	22	5	859	866	866	0.0	0:00:01
R10-3	10	5	1	799	799	799	0.0	0:00:01
R10-4	3	100	23	1266	1330	1330	0.0	0:00:03
R10-4	4	220	131	1066	1162	1162	0.0	0:00:12
R10-4	5	74	17	986	1023	1023	0.0	0:00:01
R10-4	6	39	7	910	933	933	0.0	0:00:01
R10-4	8	20	3	897	900	900	0.0	0:00:01
R10-4	10	0	1	821	821	821	0.0	0:00:01
R10-5	3	155	31	1354	1436	1436	0.0	0:00:04
R10-5	4	73	11	1077	1115	1115	0.0	0:00:01
R10-5	5	131	27	973	1023	1023	0.0	0:00:03
R10-5	6	46	13	889	941	941	0.0	0:00:01
R10-5	8	23	17	836	870	870	0.0	0:00:01
R10-5	10	0	1	789	789	789	0.0	0:00:01

Table A.15. Branch-and-Cut results, random networks (10 nodes), unit edge lengths

Graph	K	# ineq.	# B&B nodes	LB (root)	LB (final)	UB	Gap (%)	CPU time (hh:mm:ss)
R20-1	3	136	21	1533	1558	1558	0.0	0:00:19
R20-1	4	1011	2013	1355	1462	1462	0.0	0:09:53
R20-1	5	918	1289	1230	1337	1337	0.0	0:06:07
R20-1	6	610	577	1175	1247	1247	0.0	0:02:48
R20-1	8	184	21	1147	1204	1204	0.0	0:00:08
R20-1	10	229	19	1064	1104	1104	0.0	0:00:09
R20-1	16	92	21	994	1013	1013	0.0	0:00:04
R20-2	3	2032	7349	1331	1454	1454	0.0	0:50:22
R20-2	4	2200	5265	1129	1205	1205	0.0	0:31:25
R20-2	5	615	349	1051	1085	1085	0.0	0:01:59
R20-2	6	345	79	983	1014	1014	0.0	0:00:46
R20-2	8	619	525	921	959	959	0.0	0:02:10
R20-2	10	169	27	895	908	908	0.0	0:00:10
R20-2	16	73	7	861	874	874	0.0	0:00:02
R20-3	3	9282	73863	1444	1588	1675	5.5	10:00:00
R20-3	4	16332	85335	1233	1355	1355	0.0	8:53:42
R20-3	5	346	177	1158	1191	1191	0.0	0:00:45
R20-3	6	281	131	1121	1146	1146	0.0	0:00:32
R20-3	8	305	239	1080	1100	1100	0.0	0:00:42
R20-3	10	93	11	1075	1078	1078	0.0	0:00:03
R20-3	16	55	1	1051	1051	1051	0.0	0:00:01
R20-4	3	1144	3367	1640	1778	1778	0.0	0:17:35
R20-4	4	937	1333	1414	1515	1515	0.0	0:07:40
R20-4	5	219	29	1290	1307	1307	0.0	0:00:16
R20-4	6	156	17	1227	1241	1241	0.0	0:00:09
R20-4	8	220	43	1146	1167	1167	0.0	0:00:16
R20-4	10	36	3	1127	1130	1130	0.0	0:00:01
R20-4	16	23	5	1081	1084	1084	0.0	0:00:01
R20-5	3	1225	1361	1673	1765	1765	0.0	0:11:18
R20-5	4	11792	26695	1334	1471	1471	0.0	2:47:13
R20-5	5	816	543	1222	1290	1290	0.0	0:03:29
R20-5	6	383	45	1123	1188	1188	0.0	0:00:36
R20-5	8	286	19	1039	1086	1086	0.0	0:00:18
R20-5	10	617	129	1020	1069	1069	0.0	0:01:16
R20-5	16	100	1	986	986	986	0.0	0:00:01

Table A.16. Branch-and-Cut results, random networks (20 nodes), unit edge lengths

Graph	K	# ineq.	# B&B nodes	LB (root)	LB (final)	UB	Gap (%)	CPU time (hh:mm:ss)
R30-1	3	2513	35597	2087	2230	2318	3.9	10:00:00
R30-1	4	2745	21999	1729	1848	1848	0.0	3:50:34
R30-1	5	1303	2593	1593	1683	1683	0.0	0:18:50
R30-1	6	1104	1805	1532	1587	1587	0.0	0:13:42
R30-1	8	515	79	1329	1400	1400	0.0	0:01:34
R30-1	10	668	81	1256	1299	1299	0.0	0:02:09
R30-1	16	1129	227	1167	1202	1202	0.0	0:02:54
R30-2	3	8552	24765	1782	1897	2042	7.6	10:00:00
R30-2	4	13894	16313	1557	1643	1761	7.2	10:00:00
R30-2	5	15851	12943	1448	1517	1597	5.3	10:00:00
R30-2	6	5758	5725	1371	1445	1445	0.0	3:54:31
R30-2	8	1547	171	1266	1321	1321	0.0	0:17:01
R30-2	10	1558	123	1221	1266	1266	0.0	0:13:19
R30-2	16	957	207	1167	1188	1188	0.0	0:06:22
R30-3	3	3338	13913	1755	1883	1883	0.0	3:33:29
R30-3	4	21544	29705	1513	1592	1753	10.1	10:00:00
R30-3	5	16663	31873	1376	1473	1555	5.6	10:00:00
R30-3	6	5983	20351	1308	1398	1398	0.0	6:30:42
R30-3	8	1915	1981	1227	1270	1270	0.0	0:33:28
R30-3	10	7177	7163	1176	1215	1215	0.0	2:35:19
R30-3	16	1143	277	1099	1125	1125	0.0	0:05:04
R30-4	3	6965	32067	1576	1652	1804	9.2	10:00:00
R30-4	4	21088	27833	1343	1400	1564	11.7	10:00:00
R30-4	5	26318	27607	1248	1302	1445	11.0	10:00:00
R30-4	6	18505	24457	1186	1244	1321	6.2	10:00:00
R30-4	8	6991	8777	1118	1172	1172	0.0	3:37:22
R30-4	10	2756	1585	1074	1109	1109	0.0	0:39:20
R30-4	16	5850	2283	999	1035	1035	0.0	0:57:11
R30-5	3	2860	24165	2064	2214	2214	0.0	5:59:16
R30-5	4	17360	34977	1793	1892	1975	4.4	10:00:00
R30-5	5	16415	32867	1653	1749	1850	5.8	10:00:00
R30-5	6	7466	16361	1576	1670	1670	0.0	4:39:52
R30-5	8	1388	299	1473	1511	1511	0.0	0:09:11
R30-5	10	1105	171	1420	1447	1447	0.0	0:06:03
R30-5	16	2241	533	1339	1376	1376	0.0	0:10:44

Table A.17. Branch-and-Cut results, random networks (30 nodes), unit edge lengths

Graph	K	# ineq.	# B&B nodes	LB (root)	LB (final)	UB	Gap (%)	CPU time (hh:mm:ss)
R40-1	3	2198	8679	2220	2306	2482	7.6	3:00:00
R40-1	4	3932	6535	1936	2007	2187	9.0	3:00:00
R40-1	5	4590	5105	1800	1863	2018	8.3	3:00:00
R40-1	6	5229	5349	1706	1769	1875	6.0	3:00:00
R40-1	8	5033	4523	1592	1650	1710	3.6	3:00:00
R40-1	10	5773	4799	1527	1581	1619	2.4	3:00:00
R40-1	16	4569	1289	1448	1481	1481	0.0	1:13:02
R40-2	3	2650	6861	2130	2248	2460	9.4	3:00:00
R40-2	4	3675	3747	1863	1929	2092	8.4	3:00:00
R40-2	5	3920	3065	1716	1784	1937	8.6	3:00:00
R40-2	6	4313	3581	1632	1691	1787	5.7	3:00:00
R40-2	8	3835	3221	1522	1579	1588	0.6	3:00:00
R40-2	10	5056	3285	1456	1504	1517	0.9	3:00:00
R40-2	16	3552	1391	1360	1405	1405	0.0	1:03:36
R40-3	3	1708	10443	2431	2568	2584	0.6	3:00:00
R40-3	4	3221	8643	2098	2185	2360	8.0	3:00:00
R40-3	5	3587	9307	1919	2004	2087	4.1	3:00:00
R40-3	6	3953	10907	1771	1852	1929	4.2	3:00:00
R40-3	8	2591	4041	1649	1703	1703	0.0	0:54:33
R40-3	10	2083	1177	1568	1619	1619	0.0	0:20:58
R40-3	16	898	287	1494	1513	1513	0.0	0:05:51
R40-4	3	2006	8013	2258	2352	2585	9.9	3:00:00
R40-4	4	4512	8853	1956	2019	2159	6.9	3:00:00
R40-4	5	4640	8213	1790	1871	1946	4.0	3:00:00
R40-4	6	6387	8367	1703	1762	1819	3.2	3:00:00
R40-4	8	7271	8281	1562	1617	1662	2.8	3:00:00
R40-4	10	2598	1351	1495	1543	1543	0.0	0:40:59
R40-4	16	3173	2815	1415	1445	1445	0.0	0:48:26
R40-5	3	2625	9389	1872	1940	1968	1.4	3:00:00
R40-5	4	3613	7667	1711	1768	1806	2.1	3:00:00
R40-5	5	4282	6479	1623	1697	1769	4.2	3:00:00
R40-5	6	4433	7277	1571	1658	1694	2.2	3:00:00
R40-5	8	2952	1225	1465	1560	1560	0.0	0:42:43
R40-5	10	7359	4863	1299	1395	1468	5.2	3:00:00
R40-5	16	7062	3667	1178	1238	1266	2.3	3:00:00

Table A.18. Branch-and-Cut results, random networks (40 nodes), unit edge lengths

Graph	K	# ineq.	# B&B nodes	LB (root)	LB (final)	UB	Gap (%)	CPU time (hh:mm:ss)
R50-1	3	1866	3945	2208	2288	2498	9.2	3:00:00
R50-1	4	3224	4867	1923	2000	2255	12.8	3:00:00
R50-1	5	3578	4289	1786	1855	2005	8.1	3:00:00
R50-1	6	3815	4965	1703	1759	1907	8.4	3:00:00
R50-1	8	4822	3611	1579	1644	1769	7.6	3:00:00
R50-1	10	4811	2891	1522	1574	1618	2.8	3:00:00
R50-1	16	5025	3519	1441	1478	1491	0.9	3:00:00
R50-2	3	1978	4467	2460	2531	2775	9.6	3:00:00
R50-2	4	3342	3859	2151	2209	2547	15.3	3:00:00
R50-2	5	3930	4045	1999	2061	2288	11.0	3:00:00
R50-2	6	4050	3625	1894	1957	2135	9.1	3:00:00
R50-2	8	4463	2879	1774	1835	1906	3.9	3:00:00
R50-2	10	4126	3211	1713	1765	1834	3.9	3:00:00
R50-2	16	5510	3777	1621	1662	1677	0.9	3:00:00
R50-3	3	2247	4637	2378	2464	2715	10.2	3:00:00
R50-3	4	3335	3991	2068	2133	2390	12.0	3:00:00
R50-3	5	3678	3249	1945	2006	2225	10.9	3:00:00
R50-3	6	3742	4065	1860	1917	2068	7.9	3:00:00
R50-3	8	4128	3367	1762	1806	1853	2.6	3:00:00
R50-3	10	4408	3175	1703	1742	1771	1.7	3:00:00
R50-3	16	6009	2475	1627	1660	1686	1.6	3:00:00
R50-4	3	2288	5271	2341	2405	2600	8.1	3:00:00
R50-4	4	3229	4651	2015	2080	2245	7.9	3:00:00
R50-4	5	4353	3991	1868	1926	2119	10.0	3:00:00
R50-4	6	3884	3775	1784	1840	2019	9.7	3:00:00
R50-4	8	4117	3829	1672	1719	1795	4.4	3:00:00
R50-4	10	4367	3649	1597	1643	1685	2.6	3:00:00
R50-4	16	1510	177	1509	1541	1541	0.0	0:20:04
R50-5	3	2068	3683	2226	2297	2517	9.6	3:00:00
R50-5	4	2488	1313	1929	1994	2252	12.9	3:00:00
R50-5	5	2864	987	1793	1849	2004	8.4	3:00:00
R50-5	6	3867	657	1710	1755	1923	9.6	3:00:00
R50-5	8	3607	671	1606	1650	1751	6.1	3:00:00
R50-5	10	3867	795	1541	1586	1634	3.0	3:00:00
R50-5	16	4188	1503	1475	1505	1550	3.0	3:00:00

Table A.19. Branch-and-Cut results, random networks (50 nodes), unit edge lengths

A.4 HEURISTICS FOR EUCLIDEAN EDGE LENGTHS

Graph	*K*	ear-insert	cut-in-two	path-follow	stingy
B12	200	680	**675**	783	684
B12	250	**541**	558	650	**541**
B12	300	558	558	576	**541**
B12	350	558	**521**	**521**	**521**
B12	400	567	**521**	**521**	**521**
B12	450	547	521	521	**516**
B12	500	**496**	**496**	**496**	**496**
B17	200	923	933	985	**920**
B17	250	845	849	990	**789**
B17	300	823	**785**	865	789
B17	350	784	746	768	**743**
B17	400	767	746	**720**	735
B17	450	778	706	**690**	715
B17	500	758	706	**705**	715
B30	150	**1304**	X	1390	1323
B30	200	1194	1169	1309	**1157**
B30	250	1068	1003	1231	**977**
B30	300	1042	979	1218	**951**
B30	350	1014	950	1057	**910**
B30	400	1041	**950**	983	969
B30	450	951	**912**	956	919
B30	500	931	883	**879**	919
B52	150	1522	1600	1618	**1367**
B52	200	1380	1368	1526	**1283**
B52	250	1307	1291	1429	**1185**
B52	300	1345	1221	1331	**1164**
B52	350	1328	**1195**	1342	1232
B52	400	1275	1170	1232	**1158**
B52	450	1236	1170	1287	**1130**
B52	500	1257	1170	1162	**1108**
Best		3	8	7	20

Table A.20. Constructive heuristics, real applications, Euclidean edge lengths

Graph	K	ear-insert	cut-in-two	path-follow	stingy
R10-1	350	990	**906**	1178	**906**
R10-1	400	990	X	1180	**976**
R10-1	450	**923**	1102	1133	955
R10-1	500	915	947	992	**880**
R10-2	400	X	X	X	**1154**
R10-2	450	1083	X	1508	**1062**
R10-2	500	1161	X	1394	**1062**
R10-3	300	X	X	**1345**	1433
R10-3	350	**1035**	X	1172	1089
R10-3	400	**954**	1316	1057	1025
R10-3	450	**985**	1035	**985**	1025
R10-3	500	**982**	1035	985	1007
R10-4	300	X	X	X	**1445**
R10-4	350	**1217**	X	1330	1256
R10-4	400	**1147**	1230	1306	1186
R10-4	450	1177	**1023**	1042	1149
R10-4	500	1037	**1023**	1066	1142
R10-5	350	X	X	**1417**	1457
R10-5	400	**1241**	X	1353	1456
R10-5	450	1302	**1200**	1308	1292
R10-5	500	**1137**	1200	1231	1213
Best		9	4	3	7

Table A.21. Constructive heuristics, random networks (10 nodes), Euclidean edge lengths

Graph	*K*	ear-insert	cut-in-two	path-follow	stingy
R20-1	200	X	X	**1729**	1859
R20-1	250	1518	X	**1484**	1501
R20-1	300	1473	1592	1471	**1430**
R20-1	350	**1385**	1460	1406	1459
R20-1	400	1396	**1303**	1324	1416
R20-1	450	1310	**1190**	1317	1266
R20-1	500	1282	**1131**	1149	1185
R20-2	200	X	X	X	**1360**
R20-2	250	**1152**	1175	1263	1166
R20-2	300	1107	1111	1158	**1065**
R20-2	350	1115	979	1070	**974**
R20-2	400	1060	979	1066	**974**
R20-2	450	1035	979	985	**959**
R20-2	500	994	930	985	**925**
R20-3	200	X	X	X	**1512**
R20-3	250	**1244**	X	1515	1272
R20-3	300	1298	X	1417	**1266**
R20-3	350	**1161**	1446	1394	1266
R20-3	400	**1150**	1222	1295	1239
R20-3	450	**1151**	1222	1336	1228
R20-3	500	1186	**1069**	1149	1219
R20-4	200	X	X	X	**1962**
R20-4	250	1571	1772	1767	**1524**
R20-4	300	1459	1459	1494	**1442**
R20-4	350	1368	1350	1481	**1334**
R20-4	400	1373	**1228**	1389	1259
R20-4	450	1356	1228	1429	**1213**
R20-4	500	1319	1228	1235	**1213**
R20-5	300	**1359**	X	1520	1419
R20-5	350	**1275**	1363	1613	1355
R20-5	400	1314	1303	1499	**1300**
R20-5	450	1286	1191	1409	**1162**
R20-5	500	1314	1191	1219	**1125**
Best		8	5	2	18

Table A.22. Constructive heuristics, random networks (20 nodes), Euclidean edge lengths

Graph	K	ear-insert	cut-in-two	path-follow	stingy
R30-1	200	1967	X	**1899**	1963
R30-1	250	**1652**	2278	1868	1689
R30-1	300	**1567**	1821	1712	1657
R30-1	350	1584	**1541**	1704	1622
R30-1	400	**1456**	1470	1647	1464
R30-1	450	1370	**1312**	1422	1382
R30-1	500	1500	1312	**1251**	1322
R30-2	300	**1556**	1697	1893	1607
R30-2	350	1468	1512	1939	**1461**
R30-2	400	**1348**	1443	1784	1425
R30-2	450	**1348**	1443	1740	1386
R30-2	500	1394	**1375**	1457	1386
R30-3	250	**1573**	X	1824	1711
R30-3	300	**1476**	1533	1589	1526
R30-3	350	1375	1429	1547	**1250**
R30-3	400	1377	**1338**	1566	1349
R30-3	450	1368	1261	1545	**1160**
R30-3	500	1302	1261	1239	**1151**
R30-4	200	X	X	1663	**1612**
R30-4	250	**1437**	X	1563	1547
R30-4	300	**1467**	1524	1485	1484
R30-4	350	1322	**1268**	1379	1470
R30-4	400	1305	**1164**	1187	1215
R30-4	450	1321	1164	**1109**	1206
R30-4	500	1316	1164	**1107**	1223
R30-5	200	X	X	X	**2282**
R30-5	250	2163	2476	**2036**	2195
R30-5	300	**1945**	2076	2081	2097
R30-5	350	1769	1847	1916	**1668**
R30-5	400	1728	1767	1854	**1578**
R30-5	450	1610	1670	1638	**1510**
R30-5	500	1589	1612	1564	**1498**
Best		11	7	4	10

Table A.23. Constructive heuristics, random networks (30 nodes), Euclidean edge lengths

Graph	K	ear-insert	cut-in-two	path-follow	stingy
R40-1	200	X	X	X	**2549**
R40-1	250	**2195**	X	2369	2261
R40-1	300	**2056**	2185	2244	2213
R40-1	350	2016	**1855**	2089	1998
R40-1	400	1732	1855	2065	**1699**
R40-1	450	1873	**1682**	1963	1767
R40-1	500	1868	**1590**	1958	1751
R40-2	250	**1936**	2033	2092	2002
R40-2	300	1839	1780	1947	**1737**
R40-2	350	1809	1735	1830	**1592**
R40-2	400	1755	1685	1851	**1544**
R40-2	450	1639	1606	1715	**1540**
R40-2	500	1636	1550	1623	**1505**
R40-3	200	**2324**	X	2404	2421
R40-3	250	2160	2478	2218	**2146**
R40-3	300	2127	2111	2159	**1897**
R40-3	350	1902	1978	1999	**1747**
R40-3	400	1802	1896	1993	**1649**
R40-3	450	1853	1839	1875	**1622**
R40-3	500	1813	1755	1824	**1622**
R40-4	200	2182	X	2305	**2159**
R40-4	250	2031	2206	2123	**1877**
R40-4	300	1957	1884	2038	**1768**
R40-4	350	1791	1839	1871	**1664**
R40-4	400	1835	1839	1909	**1603**
R40-4	450	1770	1653	1908	**1590**
R40-4	500	1784	1621	1899	**1511**
R40-5	200	X	X	1801	**1747**
R40-5	250	**1675**	X	1677	1727
R40-5	300	**1619**	1981	1677	1712
R40-5	350	**1559**	1571	1799	1699
R40-5	400	1528	**1527**	1580	1699
R40-5	450	1480	**1459**	1532	1699
R40-5	500	1505	**1352**	1560	1699
Best		7	6	0	21

Table A.24. Constructive heuristics, random networks (40 nodes), Euclidean edge lengths

Graph	K	ear-insert	cut-in-two	path-follow	stingy
R50-1	150	X	X	X	**2367**
R50-1	200	2058	2329	2412	**2036**
R50-1	250	1980	**1929**	2262	1968
R50-1	300	1919	**1740**	2202	1946
R50-1	350	1776	**1740**	2129	1851
R50-1	400	1785	**1638**	1895	1821
R50-1	450	1718	**1611**	1889	1616
R50-1	500	1662	**1559**	1977	1630
R50-2	250	2335	2372	2608	**2267**
R50-2	300	2123	2164	2412	**1914**
R50-2	350	1945	1966	2372	**1866**
R50-2	400	2068	1966	2186	**1836**
R50-2	450	1978	1966	2168	**1824**
R50-2	500	1924	1791	2150	**1746**
R50-3	200	**2135**	X	2455	2236
R50-3	250	2161	2200	2230	**2073**
R50-3	300	**1947**	2078	2248	2013
R50-3	350	**1873**	1883	2108	1963
R50-3	400	1914	**1883**	1928	1969
R50-3	450	1936	1831	**1732**	1958
R50-3	500	1821	**1648**	1704	1854
R50-4	200	2311	2428	2408	**2183**
R50-4	250	**2085**	2271	2448	2105
R50-4	300	1995	1907	2359	**1854**
R50-4	350	1836	1865	2234	**1680**
R50-4	400	1884	1865	2048	**1729**
R50-4	450	1850	1801	2029	**1669**
R50-4	500	1890	1760	1948	**1650**
R50-5	200	**2147**	X	2248	2155
R50-5	250	1982	2080	2086	**1890**
R50-5	300	1898	1991	2083	**1770**
R50-5	350	**1715**	1789	1978	1725
R50-5	400	1788	1751	1926	**1716**
R50-5	450	**1672**	1734	1908	1686
R50-5	500	1766	**1601**	1840	1632
Best		7	9	1	18

Table A.25. Constructive heuristics, random networks (50 nodes), Euclidean edge lengths

Graph	*K*	stingy	tabu	best iteration	CPU time	guarantee (%)
B12	200	684	622	54	37	0.0
B12	250	541	541	0	30	0.0
B12	300	541	541	0	24	0.0
B12	350	521	521	0	24	0.0
B12	400	521	521	0	19	0.0
B12	450	516	516	0	16	0.0
B12	500	496	496	0	15	0.0
B17	200	920	834	5	129	0.0
B17	250	789	789	0	104	0.0
B17	300	789	726	2	77	0.0
B17	350	743	725	6	66	0.0
B17	400	735	720	184	63	0.0
B17	450	715	689	21	55	0.0
B17	500	715	689	14	52	0.0
B30	150	1323	1288	145	982	9.2
B30	200	1157	1065	381	559	0.0
B30	250	977	938	823	454	0.1
B30	300	951	898	223	350	0.0
B30	350	910	874	109	360	0.0
B30	400	969	870	306	276	0.2
B30	450	919	869	348	266	0.1
B30	500	919	844	205	263	0.0
B52	150	1367	1339	47	4090	7.2
B52	200	1283	1225	16	2538	5.2
B52	250	1185	1165	513	2280	3.6
B52	300	1164	1152	8	1677	5.1
B52	350	1232	1104	7	1909	2.0
B52	400	1158	1081	336	2173	1.4
B52	450	1130	1068	37	1627	1.4
B52	500	1108	1068	46	1578	1.4

Table A.26. Tabu Search, real applications, Euclidean edge lengths

Graph	K	stingy	tabu	best iteration	CPU time	guarantee (%)
R10-1	350	906	906	0	12	0.0
R10-1	400	976	898	3	18	0.0
R10-1	450	955	898	65	18	0.2
R10-1	500	880	854	420	18	0.0
R10-2	400	1154	1140	3	16	0.0
R10-2	450	1062	1062	0	23	0.0
R10-2	500	1062	1031	43	21	0.0
R10-3	300	1433	1269	3	15	0.0
R10-3	350	1089	1018	1	18	0.0
R10-3	400	1025	954	1	21	0.0
R10-3	450	1025	952	6	18	0.0
R10-3	500	1007	952	23	18	0.0
R10-4	300	1445	1391	1	6	0.0
R10-4	350	1256	1205	52	18	0.0
R10-4	400	1186	1117	146	18	0.0
R10-4	450	1149	980	37	18	0.0
R10-4	500	1142	980	25	17	0.0
R10-5	350	1457	1383	1	15	0.0
R10-5	400	1456	1238	1	25	0.0
R10-5	450	1292	1143	348	23	0.0
R10-5	500	1213	1072	53	21	0.0

Table A.27. Tabu Search, random networks (10 nodes), Euclidean edge lengths

Graph	K	stingy	tabu	best iteration	CPU time	guarantee (%)
R20-1	200	1859	1819	1	130	15.3
R20-1	250	1501	1445	31	136	0.0
R20-1	300	1430	1376	10	201	0.0
R20-1	350	1459	1253	175	238	0.0
R20-1	400	1416	1183	23	190	0.0
R20-1	450	1266	1144	18	180	0.0
R20-1	500	1185	1111	315	144	0.0
R20-2	200	1360	1329	191	184	0.3
R20-2	250	1166	1098	81	204	0.4
R20-2	300	1065	990	343	170	0.6
R20-2	350	974	953	133	148	0.0
R20-2	400	974	946	955	147	0.6
R20-2	450	959	932	837	129	1.4
R20-2	500	925	900	37	110	0.0
R20-3	200	1512	1449	55	165	0.0
R20-3	250	1272	1218	226	142	0.0
R20-3	300	1266	1134	47	132	0.0
R20-3	350	1266	1100	9	168	0.0
R20-3	400	1239	1100	398	177	0.0
R20-3	450	1228	1101	557	155	0.1
R20-3	500	1219	1011	562	158	0.0
R20-4	200	1962	1958	58	194	0.0
R20-4	250	1524	1460	285	221	0.4
R20-4	300	1442	1286	7	205	0.0
R20-4	350	1334	1235	5	181	0.0
R20-4	400	1259	1190	158	176	0.0
R20-4	450	1213	1164	6	141	0.0
R20-4	500	1213	1174	16	130	2.2
R20-5	300	1419	1339	17	213	1.1
R20-5	350	1355	1251	13	181	0.0
R20-5	400	1300	1213	117	213	0.2
R20-5	450	1162	1125	1	183	0.5
R20-5	500	1125	1082	84	189	0.0

Table A.28. Tabu Search, random networks (20 nodes), Euclidean edge lengths

Graph	K	stingy	tabu	best iteration	CPU time	guarantee (%)
R30-1	200	1963	1898	355	949	10.0
R30-1	250	1689	1612	734	774	9.1
R30-1	300	1657	1474	637	724	4.3
R30-1	350	1622	1366	257	539	2.9
R30-1	400	1464	1358	138	489	4.9
R30-1	450	1382	1240	64	463	0.0
R30-1	500	1322	1221	848	431	0.7
R30-2	300	1607	1507	985	731	2.0
R30-2	350	1461	1386	740	489	0.4
R30-2	400	1425	1322	995	422	0.2
R30-2	450	1386	1347	11	401	2.1
R30-2	500	1386	1302	558	412	0.5
R30-3	250	1711	1574	349	794	11.6
R30-3	300	1526	1329	34	624	0.3
R30-3	350	1250	1239	108	535	0.0
R30-3	400	1349	1198	960	422	0.0
R30-3	450	1160	1152	102	434	0.0
R30-3	500	1151	1128	72	463	0.0
R30-4	200	1612	1549	85	635	7.0
R30-4	250	1547	1327	217	837	6.2
R30-4	300	1484	1316	311	692	13.1
R30-4	350	1470	1143	810	476	0.8
R30-4	400	1215	1068	175	423	1.7
R30-4	450	1206	1044	618	362	0.0
R30-4	500	1223	1079	11	316	3.4
R30-5	200	2282	2156	63	961	4.9
R30-5	250	2195	1915	24	889	8.9
R30-5	300	2097	1750	662	949	7.0
R30-5	350	1668	1562	167	837	0.0
R30-5	400	1578	1497	13	516	0.3
R30-5	450	1510	1459	355	467	0.5
R30-5	500	1498	1425	980	367	0.1

Table A.29. Tabu Search, random networks (30 nodes), Euclidean edge lengths

Graph	K	stingy	tabu	best iteration	CPU time	guarantee (%)
R40-1	200	2549	2232	102	1336	8.0
R40-1	250	2261	2031	226	2383	12.8
R40-1	300	2213	1947	13	1969	15.4
R40-1	350	1998	1691	49	1625	4.6
R40-1	400	1699	1609	68	1269	3.3
R40-1	450	1767	1571	572	1208	2.5
R40-1	500	1751	1537	797	1044	1.1
R40-2	250	2002	1742	588	1766	5.4
R40-2	300	1737	1621	18	1596	4.0
R40-2	350	1592	1514	79	1360	1.2
R40-2	400	1544	1477	29	1280	1.2
R40-2	450	1540	1462	449	1237	2.0
R40-2	500	1505	1422	580	1046	0.4
R40-3	200	2421	2317	7	1504	14.1
R40-3	250	2146	2077	2	2091	14.1
R40-3	300	1897	1815	4	1760	7.5
R40-3	350	1747	1654	170	1387	2.1
R40-3	400	1649	1611	270	1190	1.8
R40-3	450	1622	1575	171	1025	0.9
R40-3	500	1622	1576	452	1008	2.4
R40-4	200	2159	2069	570	1821	9.2
R40-4	250	1877	1791	564	1509	4.2
R40-4	300	1768	1706	90	1778	6.0
R40-4	350	1664	1616	959	1292	4.2
R40-4	400	1603	1552	72	1364	3.3
R40-4	450	1590	1524	373	1270	3.3
R40-4	500	1511	1492	814	1003	2.3
R40-5	200	1747	1720	9	1509	5.8
R40-5	250	1727	1657	71	1330	13.9
R40-5	300	1712	1607	20	867	15.4
R40-5	350	1699	1575	570	1235	16.2
R40-5	400	1699	1546	800	1075	17.6
R40-5	450	1699	1422	447	1013	12.3
R40-5	500	1699	1291	32	1075	3.6

Table A.30. Tabu Search, random networks (40 nodes), Euclidean edge lengths

Graph	K	stingy	tabu	best iteration	CPU time	guarantee (%)
R50-1	150	2367	2250	295	4358	3.9
R50-1	200	2036	1968	64	2796	10.8
R50-1	250	1968	1880	462	2680	13.3
R50-1	300	1946	1685	144	3042	7.4
R50-1	350	1851	1595	212	2476	4.9
R50-1	400	1821	1542	8	2405	3.8
R50-1	450	1616	1531	12	2433	4.7
R50-1	500	1630	1503	28	1917	4.1
R50-2	250	2267	2200	234	3767	16.8
R50-2	300	1914	1869	40	2611	5.5
R50-2	350	1866	1811	144	3561	5.8
R50-2	400	1836	1751	382	2325	3.9
R50-2	450	1824	1723	210	2358	4.2
R50-2	500	1746	1717	119	2559	5.2
R50-3	200	2236	2053	340	3602	9.4
R50-3	250	2073	1896	876	3210	6.7
R50-3	300	2013	1847	236	2581	8.8
R50-3	350	1963	1736	721	2527	5.0
R50-3	400	1969	1707	997	2762	5.3
R50-3	450	1958	1749	547	2429	9.5
R50-3	500	1854	1652	932	2314	4.4
R50-4	200	2183	2090	730	2831	12.9
R50-4	250	2105	1822	341	3059	6.6
R50-4	300	1854	1680	123	2203	2.8
R50-4	350	1680	1624	105	2344	2.7
R50-4	400	1729	1619	96	2350	4.0
R50-4	450	1669	1601	9	2255	4.3
R50-4	500	1650	1593	4	2114	4.8
R50-5	200	2155	1960	258	4266	10.3
R50-5	250	1890	1835	674	4205	11.2
R50-5	300	1770	1683	338	3740	9.1
R50-5	350	1725	1594	176	2798	3.3
R50-5	400	1716	1630	140	2428	7.6
R50-5	450	1686	1563	440	2288	4.1
R50-5	500	1632	1547	265	2230	4.7

Table A.31. Tabu Search, random networks (50 nodes), Euclidean edge lengths

A.5 HEURISTICS FOR UNIT EDGE LENGTHS

Graph	K	stingy	tabu	best iteration	CPU time	guarantee (%)
B12	3	829	794	5	46	0.0
B12	4	693	681	92	28	0.0
B12	5	633	606	1	18	0.0
B12	6	594	568	32	13	0.0
B12	8	541	537	1	9	0.0
B12	10	521	521	0	7	0.0
B12	16	496	496	0	5	0.0
B17	3	1214	1111	48	200	1.0
B17	4	1021	971	2	169	0.5
B17	5	892	855	507	107	0.0
B17	6	852	797	855	67	0.0
B17	8	789	766	105	46	0.0
B17	10	715	715	0	34	0.0
B17	16	715	714	4	21	0.4
B30	3	1686	1506	10	1500	6.3
B30	4	1392	1306	445	867	6.4
B30	5	1266	1165	939	745	3.4
B30	6	1191	1067	778	484	1.1
B30	8	1023	957	68	259	0.1
B30	10	951	901	523	170	0.0
B30	16	919	862	967	105	0.1
B52	3	2039	1919	6	16911	14.6
B52	4	1730	1685	73	14753	15.6
B52	5	1556	1482	78	10029	9.5
B52	6	1445	1365	896	5362	5.7
B52	8	1306	1249	218	2888	3.3
B52	10	1343	1201	422	2020	3.5
B52	16	1251	1134	715	799	3.9

Table A.32. Tabu Search, real applications, unit edge lengths

Graph	*K*	stingy	tabu	best iteration	CPU time	guarantee (%)
R10-1	3	1207	1207	0	13	0.0
R10-1	4	1021	998	1	9	0.0
R10-1	5	955	901	1	6	0.0
R10-1	6	880	867	1	4	0.0
R10-1	8	825	825	0	3	0.0
R10-1	10	792	792	0	2	0.0
R10-2	3	1359	1359	0	31	0.0
R10-2	4	1117	1109	2	17	0.0
R10-2	5	1062	1062	0	12	0.0
R10-2	6	1030	1022	6	8	0.0
R10-2	8	1015	1007	4	6	0.0
R10-2	10	1001	997	14	4	0.4
R10-3	3	1383	1276	26	11	0.0
R10-3	4	1142	1057	7	7	0.0
R10-3	5	1025	954	1	5	0.0
R10-3	6	1007	931	5	4	0.0
R10-3	8	974	866	6	2	0.0
R10-3	10	956	799	8	2	0.0
R10-4	3	1445	1330	23	23	0.0
R10-4	4	1256	1162	296	12	0.0
R10-4	5	1149	1023	16	8	0.0
R10-4	6	1105	933	8	6	0.0
R10-4	8	1105	900	9	4	0.0
R10-4	10	1105	821	6	3	0.0
R10-5	3	1483	1436	17	12	0.0
R10-5	4	1157	1115	4	8	0.0
R10-5	5	1131	1023	9	6	0.0
R10-5	6	1005	941	5	4	0.0
R10-5	8	896	870	4	3	0.0
R10-5	10	831	789	4	2	0.0

Table A.33. Tabu Search, random networks (10 nodes), unit edge lengths

Graph	K	stingy	tabu	best iteration	CPU time	guarantee (%)
R20-1	3	1621	1558	4	142	0.0
R20-1	4	1548	1548	0	57	5.9
R20-1	5	1483	1454	5	47	8.8
R20-1	6	1512	1441	21	43	15.6
R20-1	8	1204	1204	0	20	0.0
R20-1	10	1113	1104	7	13	0.0
R20-1	16	1056	1013	13	8	0.0
R20-2	3	1520	1489	88	268	2.4
R20-2	4	1255	1209	121	231	0.3
R20-2	5	1107	1094	384	125	0.8
R20-2	6	1100	1019	14	88	0.5
R20-2	8	1003	961	517	59	0.2
R20-2	10	925	908	266	43	0.0
R20-2	16	900	874	276	26	0.0
R20-3	3	1717	1675	705	318	5.5
R20-3	4	1399	1365	429	186	0.7
R20-3	5	1294	1191	63	100	0.0
R20-3	6	1266	1148	802	70	0.2
R20-3	8	1232	1100	44	47	0.0
R20-3	10	1215	1078	142	35	0.0
R20-3	16	1192	1052	923	23	0.1
R20-4	3	1998	1862	15	195	4.7
R20-4	4	1567	1525	436	130	0.7
R20-4	5	1454	1307	148	95	0.0
R20-4	6	1354	1241	8	63	0.0
R20-4	8	1258	1167	165	42	0.0
R20-4	10	1198	1130	308	30	0.0
R20-4	16	1164	1084	19	18	0.0
R20-5	3	1923	1765	137	365	0.0
R20-5	4	1534	1486	60	252	1.0
R20-5	5	1376	1296	223	147	0.5
R20-5	6	1320	1195	830	107	0.6
R20-5	8	1150	1093	39	67	0.6
R20-5	10	1116	1079	1	41	0.9
R20-5	16	1056	988	783	28	0.2

Table A.34. Tabu Search, random networks (20 nodes), unit edge lengths

Graph	K	stingy	tabu	best iteration	CPU time	guarantee (%)
R30-1	3	2387	2323	89	785	4.2
R30-1	4	1879	1848	6	477	0.0
R30-1	5	1747	1707	5	349	1.4
R30-1	6	1680	1587	79	198	0.0
R30-1	8	1532	1400	515	113	0.0
R30-1	10	1401	1299	433	73	0.0
R30-1	16	1267	1202	15	46	0.0
R30-2	3	2232	2074	22	2597	9.3
R30-2	4	1861	1751	566	2110	6.6
R30-2	5	1648	1580	502	1531	4.2
R30-2	6	1546	1487	353	954	2.9
R30-2	8	1428	1335	236	591	1.1
R30-2	10	1443	1346	327	345	6.3
R30-2	16	1384	1202	169	209	1.2
R30-3	3	2089	1965	378	1022	4.4
R30-3	4	1848	1736	220	624	9.0
R30-3	5	1641	1562	955	661	6.0
R30-3	6	1492	1415	865	452	1.2
R30-3	8	1404	1280	711	254	0.8
R30-3	10	1313	1223	170	173	0.7
R30-3	16	1179	1125	79	99	0.0
R30-4	3	1832	1804	1	1319	9.2
R30-4	4	1639	1591	8	1095	13.6
R30-4	5	1582	1450	493	702	11.4
R30-4	6	1424	1327	325	486	6.7
R30-4	8	1319	1177	299	285	0.4
R30-4	10	1255	1121	149	198	1.1
R30-4	16	1222	1043	462	99	0.8
R30-5	3	2436	2234	131	886	0.9
R30-5	4	2202	1974	705	629	4.3
R30-5	5	2025	1809	773	545	3.4
R30-5	6	1843	1699	800	374	1.7
R30-5	8	1592	1518	152	225	0.5
R30-5	10	1520	1467	52	120	1.4
R30-5	16	1463	1378	715	73	0.1

Table A.35. Tabu Search, random networks (30 nodes), unit edge lengths

Graph	K	stingy	tabu	best iteration	CPU time	guarantee (%)
R40-1	3	2578	2482	4	3223	7.6
R40-1	4	2408	2182	811	1845	8.7
R40-1	5	2225	2019	1000	1413	8.4
R40-1	6	2051	1895	506	1205	7.1
R40-1	8	2030	1713	169	609	3.8
R40-1	10	1766	1668	679	473	5.5
R40-1	16	1646	1518	239	250	2.5
R40-2	3	2598	2460	6	5254	9.4
R40-2	4	2297	2078	177	2710	7.7
R40-2	5	2048	1919	860	1726	7.6
R40-2	6	1884	1789	114	1747	5.8
R40-2	8	1789	1603	50	1019	1.5
R40-2	10	1717	1514	501	662	0.7
R40-2	16	1499	1407	281	344	0.1
R40-3	3	3011	2763	42	1655	7.6
R40-3	4	2476	2359	468	1538	8.0
R40-3	5	2187	2067	898	847	3.1
R40-3	6	2076	1929	52	467	4.2
R40-3	8	1818	1711	922	268	0.5
R40-3	10	1688	1625	182	181	0.4
R40-3	16	1642	1513	91	95	0.0
R40-4	3	2634	2585	39	3323	9.9
R40-4	4	2281	2155	997	1670	6.7
R40-4	5	2032	1962	503	1098	4.9
R40-4	6	1920	1793	668	727	1.8
R40-4	8	1699	1670	34	404	3.3
R40-4	10	1709	1543	13	306	0.0
R40-4	16	1478	1451	245	156	0.4
R40-5	3	2060	1975	3	2582	1.8
R40-5	4	1893	1818	195	1733	2.8
R40-5	5	1806	1763	789	1205	3.9
R40-5	6	1762	1694	131	1176	2.2
R40-5	8	1654	1576	64	595	1.0
R40-5	10	1622	1468.	409	458	5.2
R40-5	16	1384	1286	218	202	3.9

Table A.36. Tabu Search, random networks (40 nodes), unit edge lengths

Graph	K	stingy	tabu	best iteration	CPU time	guarantee (%)
R50-1	3	3002	2914	605	10318	27.4
R50-1	4	2391	2255	414	4509	12.8
R50-1	5	2110	1995	877	2160	7.5
R50-1	6	2066	1909	51	1606	8.5
R50-1	8	1939	1737	401	996	5.7
R50-1	10	1713	1617	873	722	2.7
R50-1	16	1641	1501	14	341	1.6
R50-2	3	3234	3189	1	12481	26.0
R50-2	4	2604	2530	266	6528	14.5
R50-2	5	2406	2287	181	2989	11.0
R50-2	6	2315	2117	560	1970	8.2
R50-2	8	2114	1906	6	970	3.9
R50-2	10	1967	1832	30	825	3.8
R50-2	16	1774	1696	544	362	2.0
R50-3	3	2821	2722	185	5827	10.5
R50-3	4	2530	2399	198	4112	12.5
R50-3	5	2383	2215	496	3392	10.4
R50-3	6	2184	2076	741	2445	8.3
R50-3	8	2117	1860	951	1091	3.0
R50-3	10	1974	1805	175	839	3.6
R50-3	16	1912	1755	936	422	5.7
R50-4	3	2795	2615	81	6256	8.7
R50-4	4	2383	2245	67	3934	7.9
R50-4	5	2203	2122	55	2661	10.2
R50-4	6	2139	2009	256	1286	9.2
R50-4	8	1946	1830	209	973	6.5
R50-4	10	1870	1680	466	744	2.3
R50-4	16	1733	1580	412	344	2.5
R50-5	3	2595	2547	184	15627	10.9
R50-5	4	2309	2252	17	7872	12.9
R50-5	5	2124	2013	55	6128	8.9
R50-5	6	2081	1918	939	5522	9.3
R50-5	8	1881	1764	172	1809	6.9
R50-5	10	1767	1634	236	1569	3.0
R50-5	16	1632	1544	200	770	2.6

Table A.37. Tabu Search, random networks (50 nodes), unit edge lengths

References

[1] E. Aarts and J.K. Lenstra, editors. *Local Search in Combinatorial Optimization*. John Wiley & Sons Ltd., 1997.

[2] A. Agrawal, P.N. Klein, and R. Ravi. When trees collide: an approximation algorithm for the generalized Steiner problem on networks. *SIAM Journal on Computing*, 24(3):440–456, 1995.

[3] A.V. Aho, J.E. Hopcroft, and J.D. Ullman. *The design and analysis of computer algorithms*. Addison-Wesley, 1974.

[4] R.K. Ahuja, T.L. Magnanti, and J.B. Orlin. *Network flows, theory, algorithms and applications*. Prentice Hall, Englewood Cliffs, NJ 07632, 1993.

[5] A. Balakrishnan, T.L. Magnanti, A. Shulman, and R.T. Wong. Models for planning capacity expansion in local access telecommunication networks. *Annals of Operations Research*, 33:239–284, 1991.

[6] F. Barahona. Separating from the dominant of the spanning tree polytope. *Op. Research Letters*, 12:201–203, 1992.

[7] R.E. Bellman. On a routing problem. *Q. Appl. Math.*, 16:87–90, 1958.

[8] C. Berge. *Graphs and hypergraphs*. North-Holland, Amsterdam, 1973.

[9] D. Bienstock, E.F. Brickell, and C.L. Monma. On the structure of minimum-weight k-connected spanning networks. *SIAM J. Discrete Math.*, 3(3):320–329, 1990.

[10] D. Bienstock, S. Chopra, O. Günlük, and C.-Y. Tsai. Minimum cost capacity installation for multicommodity network flows. *Mathematical Programming*, 81(2):177–199, 1998.

[11] J.A. Bondy and U.S.R. Murty. *Graph theory with applications.* Universities Press, Belfast, 1976.

[12] S.C. Boyd and T. Hao. An integer polytope related to the design of survivable communication networks. *SIAM J. Discrete Math.*, 6(4):612–630, 1993.

[13] S.C. Boyd and F. Zhang. Transforming clique tree inequalities to induce facets for the 2-edge connected polytope. Technical Report TR-94-13, Departement of Computer Science, University of Ottawa, 1994.

[14] G.-R. Cai and Y.-G. Sun. The minimum augmentation of any graph to a k-edge-connected graph. *Networks*, 19:151–172, 1989.

[15] R.H. Cardwell, C.L. Monma, and T.H. Wu. Computer-Aided Design Procedure for Survivable Fiber Optic Networks. *IEEE Journal on Selected Areas in Communications*, 7(8):1188–1197, 1989.

[16] S. Chopra. The k-edge-connected spanning subgraph polyhedron. *SIAM J. Discrete Math.*, 7(2):245–259, 1994.

[17] W. Chou and H. Frank. Survivable communication networks and the terminal capacity matrix. *IEEE Transactions on Circuit Theory*, CT-17:192–197, 1970.

[18] N. Christofides. Worst-case analysis of a new heuristic for the traveling salesman problem. Technical Report 388, Graduate School of Industrial Administration, Carnegie-Mellon University, Pittsburgh, PA, 1976.

[19] N. Christofides and C.A. Whitlock. Network synthesis with connectivity constraints — a survey. In J.P. Brans, editor, *Operational Research '81*, pages 705–723. North-Holland Publishing Company, 1981.

[20] L.W. Clarke and G. Anandalingam. A bootstrap heuristic for designing minimum cost survivable networks. *Computers & Operations Research*, 22:921–924, 1995.

[21] L.W. Clarke and G. Anandalingam. Heuristics for designing minimum cost survivable telecommunication networks. *IEEE Transactions on Systems, Man and Cybernetics*, 26:856–862, 1996.

[22] G. Cornuéjols, F. Fonlupt, and D. Naddef. The traveling salesman problem on a graph and some related integer polyhedra. *Mathematical Programming*, 33:1–27, 1985.

[23] C.R. Coullard, A. Rais, R.R. Rardin, and D.K. Wagner. The 2-connected-Steiner-subgraph polytope for series-parallel graphs. Technical Report CC-91-32, Purdue University, 1991.

[24] C.R. Coullard, A. Rais, R.R. Rardin, and D.K. Wagner. The dominant of the 2-connected-Steiner-subgraph polytope for w_4-free graphs. Technical Report CC-91-34, Purdue University, 1991. To appear in *Discrete Applied Mathematics*.

[25] C.R. Coullard, A. Rais, D.K. Wagner, and R.L. Rardin. Linear-time algorithms for the 2-connected Steiner subgraph problem on special classes of graphs. *Networks*, 23, 1993.

[26] W.H. Cunningham. Optimal attack and reinforcement of a network. *Journal of ACM*, 32:549–561, 1985.

[27] A. De Jongh. *Uncapacitated network design with bifurcated routing*. PhD thesis, Université Libre de Bruxelles, 1998.

[28] M. Didi Biha and A.R. Mahjoub. k-edge connected polyhedra on series-parallel graphs. *Operations Research Letters*, 19:71–78, 1996.

[29] E.W. Dijkstra. A note on two problems in connection with graphs. *Numer. Math.*, 1:269–271, 1959.

[30] K.P. Eswaran and R.E. Tarjan. Augmentation problems. *SIAM Journal on Computing*, 5:653–665, 1976.

[31] B. Fortz, M. Labbé, and F. Maffioli. Two-connected networks with bounded meshes. Technical Report IS-MG 96/8, Université Libre de Bruxelles, CP 210/01, B-1050 Bruxelles, Belgique, 1996. To appear in *Operations Research*.

[32] B. Fortz, M. Labbé, and F. Maffioli. Methods for designing reliable networks with bounded meshes. In V. Ramaswami and P.E. Wirth, editors, *Teletraffic Contributions for the Information Age*, volume 2a of *Teletraffic Science and Engineering*, pages 341–350. Elsevier, 1997.

[33] A. Frank. Augmenting graphs to meet edge-connectivity requirements. *SIAM J. on Discrete Mathematics*, 5(1):22–53, 1992.

[34] H. Frank and W. Chou. Connectivity considerations in the design of survivable networks. *IEEE Transactions on Circuit Theory*, CT-17:486–490, 1970.

[35] G.N. Frederickson and J. Já Já. Approximation algorithms for several graph augmentation problems. *SIAM Journal on Computing*, 10(2):270–283, 1981.

[36] G.N. Frederickson and J. Já Já. On the relationship between the biconnectivity augmentation and traveling salesman problems. *Theoretical Computer Science*, 19:189–201, 1982.

[37] D.R. Fulkerson and L.S. Shapley. Minimal k-arc connected graphs. *Networks*, 1:91–98, 1971.

[38] H.N. Gabow, M.X. Goemans, and D.P. Williamson. An efficient approximation algorithm for the survivable network design problem. In *Proceedings of the Third MPS Conference on Integer Programming and Combinatorial Optimization*, pages 57–74, 1993.

[39] M.R. Garey and D.S. Johnson. *Computers and intractability: A guide to the theory of NP-completeness.* Freeman, San Francisco, 1979.

[40] F. Glover. Future paths for integer programming and links to artificial intelligence. *Computers & Operations Research*, 13:533–549, 1986.

[41] F. Glover and M. Laguna. *Tabu Search.* Kluwer Academic Publishers, 1997.

[42] M.X. Goemans. Arborescence polytopes for series-parallel graphs. *Discrete Applied Mathematics*, 51:277–289, 1994.

[43] M.X. Goemans and D.J. Bertsimas. Survivable networks, lp relaxations and the parsimonious property. *Mathematical Programming*, 60:145–166, 1993.

[44] M.X. Goemans and D.P. Williamson. A general approximation technique for constrained forest problems. *SIAM Journal on Computing*, 24:296–317, 1995.

[45] M.X. Goemans and D.P. Williamson. The primal-dual method for approximation algorithms and its application to network design problems. In D. S. Hochbaum, editor, *Approximation Algorithms for NP-Hard Problems*. PWS Publishing Company, 1996.

[46] R.E. Gomory and T.C. Hu. Multi-terminal network flows. *SIAM J. Appl. Math.*, 9:551–570, 1961.

[47] M. Grötschel and C.L. Monma. Integer polyhedra arising from certain design problems with connectivity constraints. *SIAM J. Discrete Math.*, 3:502–523, 1990.

[48] M. Grötschel, C.L. Monma, and M. Stoer. Computational results with a cutting plane algorithm for designing communication networks with low-connectivity constraints. *Operations Research*, 40(2):309–330, 1992.

[49] M. Grötschel, C.L. Monma, and M. Stoer. Facets for polyhedra arising in the design of communication networks with low-connectivity constraints. *SIAM J. Optimization*, 2(3):474–504, 1992.

[50] M. Grötschel, C.L. Monma, and M. Stoer. A polyhedral approach to network connectivity problems. Extended Abstract, 1993.

[51] M. Grötschel, C.L. Monma, and M. Stoer. *Design of Survivable Networks*, volume 7 on Network models of *Handbooks in OR/MS*, chapter 10, pages 617–672. North-Holland, 1995.

[52] M. Grötschel, C.L. Monma, and M. Stoer. Polyhedral and computational investigations for designing communication networks with high survivability requirements. *Operations Research*, 43(6):1012–1024, 1995.

[53] F. Harary. The maximum connectivity of a graph. In *Proceedings of the National Academy of Sciences*, volume 48, pages 1142–1146, USA, 1962.

[54] A. Hertz, E. Taillard, and D. de Werra. Tabu search. In E. Aarts and J.K. Lenstra, editors, *Local Search in Combinatorial Optimization*, pages 121–136. John Wiley & Sons Ltd., 1997.

[55] T.-S. Hsu and V. Ramachandran. A linear time algorithm for triconnectivity augmentation. In *Proc. 32nd Annual IEEE Symposium on Foundations of Computer Science*, pages 548–559, 1991.

[56] T.-S. Hsu and V. Ramachandran. On finding a minimum augmentation to biconnect a graph. *SIAM Journal on Computing*, 22:889–891, 1993.

[57] S. Khuller. Approximation algorithms for finding highly connected subgraphs. In D. S. Hochbaum, editor, *Approximation Algorithms for NP-Hard Problems*. PWS Publishing Company, 1996.

[58] S. Khuller and R. Thurimella. Approximation algorithms for graph augmentation. *Journal of Algorithms*, 14(2):214–225, 1993.

[59] S. Khuller and U. Vishkin. Biconnectivity approximations and graph carvings. *Journal of the ACM*, 41(2):214–235, 1994.

[60] C-W. Ko and C.L. Monma. Heuristics methods for designing highly survivable communication networks. Technical report, Bellcore, 1989.

[61] J.B. Kruskal. On the shortest spanning subtree of a graph and the traveling salesman problem. *Proc. Amer. Math. Soc.*, 7:48–50, 1956.

[62] E.L. Lawler. *Combinatorial Optimization : Networks and Matroids*. Holt, Rinehart and Wilson, New-York, 1976.

[63] E.L. Lawler, J.K. Lenstra, A.H.G. Rinnooy Kan, and D.B. Shmoys, editors. *The Traveling Salesman Problem*. John Wiley & Sons, New-York, 1985.

[64] N. Maculan. The Steiner problem in graphs. *Annals of Discrete Mathematics*, 31:185–221, 1987.

[65] F. Maffioli and P. Värbrand. The constrained two-connected subgraph problem. Technical Report 94.013, Politecnico di Milano, 1994.

[66] A.R. Mahjoub. Two-edge connected spanning subgraphs and polyhedra. *Mathematical Programming*, 64:199–208, 1994.

[67] K. Menger. Zur allgemeinen kurventheorie. *Fundamenta Mathematicae*, 10:96–115, 1927.

[68] C.L. Monma, B.S. Munson, and W.R. Pulleyblank. Minimum-weight two-connected spanning networks. *Mathematical Programming*, 46:153–171, 1990.

[69] C.L. Monma and D.F. Shallcross. Methods for designing communications networks with certain two-connected survivability constraints. *Operations Research*, 37(4):531–541, 1989.

[70] D. Naor, D. Gusfield, and Ch. Martel. A fast algorithm for optimally increasing the edge-connectivity. In *Proceedings of the Foundation of Computer Science '90*, pages 698–707, St. Louis, 1990.

[71] G.L. Nemhauser and L.A. Wolsey. *Integer and combinatorial optimization*. Wiley-Interscience series in discrete mathematics and optimization. Wiley, 1988.

[72] R.C. Prim. Shortest connection networks and some generalizations. *Bell System Tech. J.*, 36:1389–1401, 1957.

[73] A. Rosenthal and A. Goldner. Smallest augmentation to biconnect a graph. *SIAM Journal on Computing*, 6:55–66, 1977.

[74] K. Steiglitz, P. Weiner, and D.J. Kleitman. The design of minimum-cost survivable networks. *IEEE Transactions on Circuit Theory*, CT-16:455–460, 1969.

[75] M. Stoer. *Design of Survivable Networks*, volume 1531 of *Lecture Notes in Mathematics*. Springer-Verlag, 1992.

[76] J.W. Suurballe. Disjoint paths in a network. *Networks*, 4:125–145, 1974.

[77] J.W. Suurballe and R.E. Tarjan. A quick method for finding shortest pairs of disjoint paths. *Networks*, 14:325–336, 1984.

[78] S. Thienel. *ABACUS—A Branch-And-Cut System*. PhD thesis, Universität zu Köln, 1995.

[79] S. Thienel. *ABACUS—A Branch-And-Cut System, Version 2.0, Users's Guide and Reference Manual*. Universität zu Köln, 1997.

[80] S. Ueno, Y. Kajitani, and H. Wada. Minimum augmentation of a tree to a k-edge-connected graph. *Networks*, 18:19–25, 1988.

[81] J.A. Wald and C.J. Colbourn. Steiner trees, partial 2-trees, and minimum ifi networks. *Networks*, 13:159–167, 1983.

[82] T. Watanabe and A. Nakamura. Edge-connectivity augmentation problems. *Computer and System Sciences*, 35:96–144, 1987.

[83] D.P. Williamson, M.X. Goemans, M. Mihail, and V.V. Vazirani. A primal-dual approximation algorithm for generalized Steiner network problems. *Combinatorica*, 15:435–454, 1995.

[84] P. Winter. Generalized Steiner problem in halin networks. In *Proc. 12th International Symposium on Mathematical Programming*. MIT, 1985.

[85] P. Winter. Generalized Steiner problem in outerplanar networks. *BIT*, 25:485–496, 1985.

[86] P. Winter. Generalized Steiner problem in series-parallel networks. *Journal of Algorithms*, 7:549–566, 1986.

[87] P. Winter. Topological network synthesis. In B. Simeone, editor, *Combinatorial Optimization - Como 1986*, volume 1403 of *Lecture Notes in Mathematics*, pages 282–303. Springer-Verlag, 1986.

[88] P. Winter. Steiner problems in networks : a survey. *Networks*, 17:129–167, 1987.

Index

ABACUS, 108, 112, 125
algorithm, 8–23
  ABACUS, 108, 112, 125
  Barahona, 106, 107
  branch-and-cut, 3, 4, 19, 26, 62, 91–113, 125, 127, 138, 143, 147
  depth-first search, 92, 94
  Dijkstra, 95
  Gomory-Hu, 95, 105
  heuristics, 115–122, 125
  separation, 4, 95–107
  Tabu Search, 119–122
  two-connectivity, 94
articulation point, 8, 47, 50, 75, 89, 94, 106

branch-and-bound, 21, 129
  tree, 112, 113, 127, 129
branch-and-cut, 3, 4, 19, 21, 22, 26, 62, 91, 107, 108, 110–113, 115, 125, 127, 129, 130, 132, 133, 135, 137, 138, 143, 147, 159–164, 166–171

connectivity, 3, 7, 9–22, 25, 26
  k-edge-connectivity, 7, 10, 16, 18
  k-node-connectivity, 7, 10, 17, 18, 22
  two-edge-connectivity, 16, 20, 40, 46, 72, 106
  two-node-connectivity, 3, 4, 16, 20, 27, 37, 40, 42–54, 56–58, 63–66, 74, 75, 77–79, 88, 92, 94, 96, 100, 101, 104, 105, 114–116, 119, 123, 147

convex hull, 8, 36
cut, 6, 8, 41, 42, 44, 47, 51, 65, 88, 89, 103, 104, 112
  constraints, 11, 21, 22, 26, 27, 30–32, 40–47, 51, 54, 105, 114, 127
  minimum, 21, 95, 105, 106
  node-cut constraints, 11, 21, 22, 27, 30–32, 46–50, 100, 127
cut-in-two heuristic, 116
cycle, 4, 7, 18, 25, 27–31, 37, 44, 45, 58–60, 62–66, 71–76, 79, 82, 92, 94, 98, 104, 109, 116–118, 129, 147
  boundary, 75
  fundamental, 70–74
  Hamiltonian, 3, 7, 12, 17, 19, 26, 77–79, 81, 116, 118, 147
  matrix, 70, 72–74
  space, 70–72, 75
cyclomatic inequalities, 81–88, 107, 114

degree, 6, 79, 93, 96, 106, 117
  constraints, 110
depth-first search, 92, 94
dimension, 8, 22, 35, 37–39, 42, 44, 55, 71

ear-inserting heuristic, 18
edge-disjoint paths, 7–10

facet, 4, 8, 22, 35–58, 63, 64, 83, 84, 86, 88, 89, 100, 101, 113, 114

failure, 3, 4, 9, 26, 27, 147
flow, 3, 5, 26, 27, 31–33, 59, 60, 97, 98, 100, 147
    maximum, 95
    minimum cost, 60, 97–99

Gomory-Hu algorithm, 21, 95, 105
Gomory-Hu tree, 105
graph, 5–8, 10, 12–22, 36–44, 46–51, 53–55, 57, 59, 63–65, 70, 71, 73, 77, 78, 82, 83, 88, 92–97, 99, 101, 102, 105, 117–119, 123, 126, 129, 150
    complete, 12, 13, 16, 30, 83, 89, 102, 104, 109, 118, 120, 126, 150
    ring-cut, 88, 89, 102–105, 112, 114
    simple, 6

Hamiltonian cycle, 3, 7, 12, 17, 19, 26, 77–79, 81, 116, 118, 147
heuristics, 4, 11, 16–20, 98, 100, 102, 105, 111, 114–122, 125, 138–143, 147, 172–185
    cut-in-two, 116
    ear-inserting, 18, 116
    path-following, 117–119
    stingy, 119
    Tabu Search, 111, 119–122, 127, 129, 148
    two-trees, 18

independent set, 6, 53–55, 57, 89, 102–105, 114
inequalities, 111, 114
    cut, 11, 21, 22, 26, 27, 30–32, 40–47, 51, 54, 105, 127
    cyclomatic, 81–88, 107
    metric, 59–62, 95–101
    node-cut, 11, 21, 22, 27, 30–32, 46–50, 100, 127
    node-partition, 63–65, 106, 127
    ring-cover, 58–59, 61, 62, 98, 100
    ring-cut, 51–58, 88, 92, 104, 105, 112
    weighted partition, 65–68, 106
integer linear program, 4, 5, 11, 26, 27, 31, 147

linear program, 21, 31, 59, 110
    integer, 4, 5, 11, 26, 27, 31, 147
    linear relaxation, 19, 21, 61, 98
linear relaxation, 19, 21, 61, 98
long-term planning, 2, 6, 26

maximum flow, 95
metric inequalities, 59–62, 95–101, 114
mid-term planning, 2
minimum cost flow, 60, 97–99

network, 1–5, 7, 9, 11, 26, 28, 71–75, 91, 112, 125–143, 147, 149–185
    backbone, 1–3, 5, 6, 8, 26, 147
    connected, 7, 22, 63, 106
    design, 3, 9, 11–13, 15, 17, 19–22, 40, 46
    survivable, 2, 3, 5–23, 25, 36, 147
    topology, 2, 3
    two-connected, 3, 4, 12, 16–20, 23, 25–33, 36, 40, 46, 49, 63, 74, 75, 78, 100, 116, 118, 127, 147
    two-edge-connected, 71, 106
node-cut constraints, 11, 21, 22, 27, 30–32, 46–50, 100, 127
node-disjoint paths, 3, 7–10, 27, 31, 79
node-partition inequalities, 63–65, 106, 114, 127
NP-complete, 7, 69, 70, 77, 102, 114
NP-hard, 4, 12, 13, 16, 20, 105

path, 7, 10, 18, 27, 45, 56, 64, 75, 76, 79, 82, 92, 98, 100
    edge-disjoint, 7–10
    node-disjoint, 7–10, 27, 31, 79
    shortest, 7, 13, 92, 95
path-following heuristic, 117–119
polyhedron, 8
polytope, 15, 16, 21, 22, 35, 36, 38–40, 42, 46, 63, 106
projection, 36, 58, 59

ring, 4
    bounded, 3, 25–33, 36–40, 69, 147
    constraints, 27, 28, 32, 36, 40, 49, 58, 78, 97, 115, 116, 127
    self-healing, 4, 27, 69
    triangular, 69, 88–89
ring-cover inequalities, 58–59, 61, 62, 98, 100, 114
ring-cut graph, 42–46, 51–58, 88, 89, 102–105, 112, 114
ring-cut inequalities, 51–58, 88, 92, 104, 105, 112, 114

self-healing ring, 4, 27, 69
separation, 4, 59, 62, 91, 92, 95–107, 111–114
spanning tree, 13, 18, 64, 65, 70, 73
Steiner subgraph, 16
Steiner tree, 12, 13, 15, 17
stingy heuristic, 119

survivability, 2, 3, 6, 8–10, 12, 22, 26

Tabu Search, 111, 119–122, 127, 129, 148
tree, 2, 3, 7, 45, 64, 93, 94
    branch-and-bound, 112, 113, 127, 129
    Gomory-Hu, 105
    spanning, 13, 18, 64, 65, 70, 73
    Steiner, 12, 13, 15, 17
triangle inequality, 7, 109, 118
triangular rings, 69, 88–89
trivial inequalities, 38, 39
two-trees heuristic, 18

vertex, 8, 36

weighted partition inequalities, 65–68, 106, 114

MIX
Papier aus verantwortungsvollen Quellen
Paper from responsible sources
FSC® C105338

If you have any concerns about our products,
you can contact us on
ProductSafety@springernature.com

In case Publisher is established outside the EU,
the EU authorized representative is:
Springer Nature Customer Service Center GmbH
Europaplatz 3, 69115 Heidelberg, Germany

Printed by Libri Plureos GmbH
in Hamburg, Germany